6G Multimedia Communications

6G Multimedia Communications

Analysis, Design, and Optimization

Nguyen-Son Vo
Duy Tan University, Vietnam

Trung Q. Duong
Memorial University, Canada

Trang Hoang
Ho Chi Minh City University of Technology-VNU HCM, Vietnam

IEEE Press
WILEY

Published by John Wiley & Sons, Inc., Hoboken, New Jersey.

Library of Congress Cataloging-in-Publication Data has been applied for:

Hardback ISBN: 9781394232949
ePDF ISBN: 9781394232963
ePUB ISBN: 9781394232956
oBook ISBN: 9781394232970

Cover Design: Wiley
Cover Image: Flavio Coelho/Getty Images

Contents

About the Authors

Nguyen-Son Vo is with the Institute of Fundamental and Applied Sciences, Duy Tan University, Vietnam.

Trung Q. Duong is a Canada Excellence Research Chair and a Full Professor at Memorial University (MUN), Canada and also an adjunct Professor at Queen's University Belfast (QUB), UK.

Trang Hoang is an Associate Professor at Ho Chi Minh City University of Technology-VNU HCM, Vietnam.

Preface

Every decade brings a groundbreaking transformation—a new generation of communications technologies that shapes many aspects of human life, from daily, social, economic activities to health care, education, and entertainment.

With each generation, network operators and Internet service providers are faced with escalating challenges driven by the exponential growth in the number of users who require diverse applications and services with higher quality of experience under limited system resources. These challenges become more critical in the era of sixth-generation (6G) networks, where a massive number of mobile users demand ultra-high data-rate multimedia applications and services for advanced use cases such as holography and extended reality, to name but a few. While 6G multimedia applications and services aim to deliver more immersive and realistic experiences, they impose substantial burdens in terms of energy/spectrum efficiency, computational complexity, and potential network congestion.

In this context, the development of 6G multimedia communications follows two complementary directions: technological innovations and software-defined techniques. The former focuses on disruptive hardware and infrastructure developments, such as terahertz communications, reconfigurable intelligent surfaces, satellite–terrestrial integration, massive multiple-input multiple-output, optical wireless communications, and integrated sensing and communications.

However, such technological advances usually entail significant capital investment, longer deployment time, and higher operational costs. In contrast, the latter offers a more flexible and cost-effective approach. By leveraging available system resources, artificial intelligence-based learning algorithms, and network virtualization, software-defined techniques can be dynamically implemented through edge computing, multicasting, device-to-device communications, and caching. These techniques can be further integrated with emerging solutions such as resource sharing, power allocation, scheduling, adaptive bitrate encoding, cross-layer design, and cooperative transmission to meet the stringent requirements of 6G multimedia communications.

This book focuses on software-defined techniques that leverage existing hardware, network architectures, system resources, and emerging solutions to enable the analysis, design, and optimization of 6G multimedia communications.

These techniques are expected to support various types of 6G multimedia applications and services simultaneously, provide mobile users with highly immersive quality of experience, and bridge the gap between operating systems and multimedia communications. In particular, *6G Multimedia Communications: Analysis, Design, and Optimization* covers the following topics:

- A comprehensive overview of theoretical foundations and emerging concepts in multimedia applications and services and 6G multimedia communications.
- Practical software-defined techniques, including edge computing, multicasting, device-to-device communications, and caching, to support diverse 6G multimedia applications and services.
- Advanced solutions, such as resource sharing, power allocation, scheduling, adaptive bitrate encoding, cross-layer design, and cooperative transmission, leverage artificial intelligence-based learning algorithms to optimize multimedia delivery performance and system resources.
- Strategies to enhance mobile users' quality of experience while delivering substantial benefits to content providers and Internet service providers.

Nguyen-Son Vo, Trung Q. Duong, and Trang Hoang

Acronyms

6G	Sixth Generation
ABR	Adaptive Bitrate
AI	Artificial Intelligence
AR	Augmented Reality
AVC	Advanced Video Coding
AWGN	Additive White Gaussian Noise
BS	Base Station
CDN	Content Delivery Networks
CLD	Cross-layer Design
D2D	Device-to-device
eMBB	Enhanced Mobile Broadband
FBS	Femto-cell Base Station
FEC	Forward Error Correction
GAs	Genetic Algorithms
IBM	Indian Buffet Model
IoT	Internet of Things
MAS	Multimedia Applications and Services
MBS	Macro Base Station
MHMP	Multi-hop Multi-path
MSE	Mean Squared Error
MTC	Machine-type Communications
mMTC	Massive MTC
MU	Mobile User
NOMA	Non-orthogonal Multiple Access
PSNR	Peak Signal-to-noise Ratio
QoE	Quality of Experience
QoS	Quality of Service
RDM	Rate–distortion Model
RS	Reed–Solomon

SBS	Small-cell Base Station
SHVC	Scalable extensions of High Efficiency Video Coding
SINR	Signal-to-interference-plus-noise Ratio
SNR	Signal-to-noise Ratio
UAV	Unmanned Aerial Vehicles
UDN	Ultra-dense Networks
URLLC	Ultra-reliable and Low-latency Communications
VANET	Vehicular Ad hoc Networks
VAS	Video Applications and Services
VR	Virtual Reality
XR	Extended Reality

1

An Overview of Multimedia Communications

1.1 Advanced Applications and Services in 6G Networks

Over the span of 40 years of mobile communications development, various applications and services have transformed our work and life: from 2G simple voice and short messaging service to 3G broadband data, 4G machine-type communications (MTC) and videos, and 5G industrial Internet of Things (IoT) and interactive videos. The successful launch of 5G networks and technologies has introduced a disruptive revolution. In the fourth industrial revolution, 5G networks have focused on the three advanced applications and services of enhanced mobile broadband (eMBB), massive MTC (mMTC), and ultra-reliable and low-latency communications (URLLCs).

Immediately, 6G networks and technologies have begun to address what 5G could not achieve: enabling immersive communications and driving digital transformation across various sectors to bridge the physical and virtual worlds. 6G is anticipated to emerge around 2030, significantly advancing beyond 5G in peak performance with eMBB+, mMTC+, and URLLC+, while further integrating sensing, digital twin, and artificial intelligence technologies. At the same time, numerous mobile users (MUs) demand advanced applications and services for greater visual and realistic sensations. In this context, a key set of 6G performance indicators and goals, such as connected devices per square kilometer, latency, and reliability, are carried out for MUs to achieve ultimate experience in all advanced applications and services. Especially, leveraging frequency range from 100 GHz to 1 THz is expected to enable a data rate up to 1 Tbps and a latency of 0.1 ms, respectively, representing fifty times higher and ten times lower compared to those in 5G.

In any development phase of mobile communications, there is a crucial need for groundbreaking techniques that can serve the massive number of MUs of advanced applications and services. 6G techniques are expected to achieve a seamless integration between devices, machines, sensing and communications, optimization and computing algorithms, and management to satisfy quality of service (QoS) and quality of experience (QoE) requirements. Based on this, interconnected IoT systems in smart cities are developed to provide the MUs with access to various advanced applications and services of social media, e-health, e-education, intelligent transportation, public safety, city government, smart businesses, and entertainment, for a better quality of life.

1.2 MAS in 6G Networks

By the year 2033, there will be about 32 billion IoT devices worldwide [1], causing the global mobile data traffic to increase exponentially. Figure 1.1 depicts 6G MAS, including (i) monitoring, automation, holography, haptics (MAH), (ii) infotainment, 360° videos, video gaming, 4K/8K/12K video streaming (XV), and (iii) virtual reality/augmented reality/mixed reality, namely extended reality (XR). The data generated by applications and services within these three categories accounts for approximately 79% of the mobile data traffic [2] and is the main focus of Internet service and content providers. Unlike conventional MAS, 6G MAS

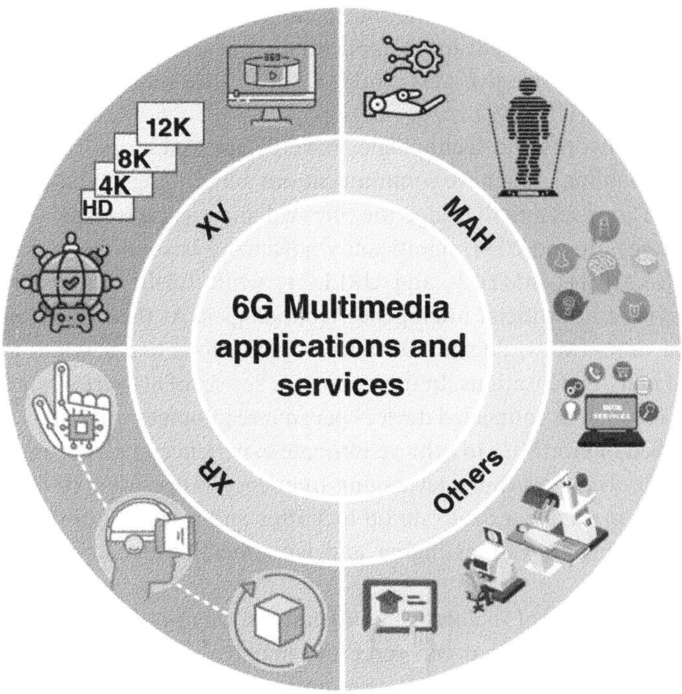

Figure 1.1 6G multimedia applications and services.

require emerging 6G-based commercial platforms to provide MUs with greater immersion. In MAH, a proliferation of multimedia collaboration among different functional areas is produced by multimodal sensors and smart objects. XV consumes much larger system bandwidth and needs quick response to passive MUs, while XR devices of active MUs are further equipped with local video decoding, post-processing, and rendering [3]. Particularly, virtual reality, a computer-based simulation for 3D experience using reality headsets add-on, can bring MUs to an imaginary world to interact with virtual objects and explore the environments through the realistic sensations of all five senses. For a real-world setting, in augmented reality, the virtual elements are augmented by various computer-generated sensor inputs and added to the existing physical world for interacting and exploring. Finally, as a hybrid reality, mixed reality merges virtual reality and augmented reality to establish a real-time world in which the artificial, physical, and biological objects can interact with each other in real time.

6G MAS also finds their role in other aspects of life, from health care and education, tourism and culture, marketing and retail, to architecture, construction, and system design. 6G MAS needs a huge pool of resources, powerful computing and processing capabilities, and scalable infrastructures that are extremely challenging for conventional multimedia communications. On the other hand, the satisfaction of MUs with 6G MAS is represented not only by QoS with simple throughput and delay, but also by QoE, which is evaluated at all times, i.e., when an MU sends a request for joining, waits for approval (the first bits received), enjoys the MAS, and interacts with others until it leaves.

The rapid growth of 6G MAS and the rising demand for high QoE from a massive number of MUs, despite limited system resources, have stimulated an urgent need for analyzing, designing, and optimizing 6G multimedia communications. To this end, emerging 6G techniques that can be flexibly implemented and integrated into new technologies of sensing and artificial intelligence, digital twin, brain-computer interfaces, autonomous things, low- or very low-earth orbit satellites, e.g., unmanned aerial vehicles (UAV), and THz bands, are studied. These techniques are expected to support more types of 6G MAS at once, providing vastly immersive QoE, and bridging the gap between the operating systems, multimedia communications, and the MUs' ubiquity in real time.

1.3 Techniques for 6G Multimedia Communications

6G multimedia communications place a higher priority on designing software-defined techniques and utilizing the available system resources. Recently, edge techniques have drawn significant attention from both academic and industrial sectors as a most efficient method for multimedia communications that can benefit Internet service and content providers and satisfy MUs' huge demand for MAS

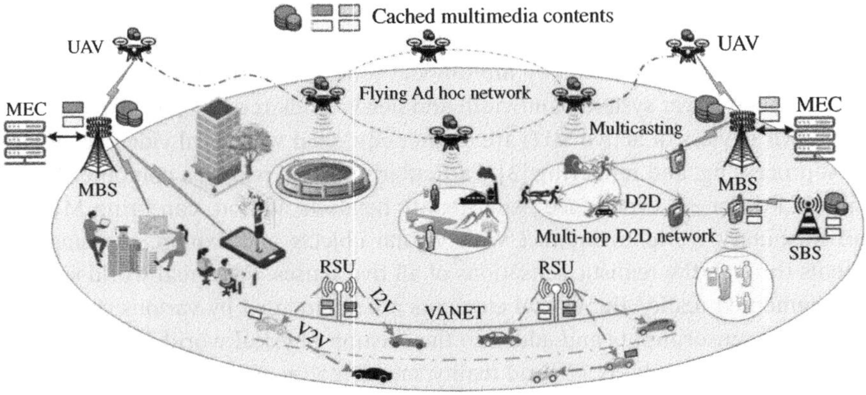

Figure 1.2 6G multimedia communications.

over 5G and beyond networks. In what follows, we will review and discuss these techniques, e.g., edge computing, multicasting, device-to-device (D2D) communications, and caching, combined with emerging solutions of resource sharing, power allocation, scheduling, adaptive bitrate (ABR) encoding, cross-layer design, and cooperative transmission, as shown in Figure 1.2.

1.3.1 Edge Computing Assisted Multimedia Communications

In Li et al. [4], the authors proposed a mobile edge computing paradigm to facilitate the bitrate adaptation and video transcoding for MAS with the help of local edge servers. Depending on the requirement and transmission condition of an MU, the corresponding edge server and encoding bitrate are selected. Time-consuming transcoding tasks are then carefully scheduled, considering edge computing capacity allocation to minimize the delay in MAS. To improve the performance of mobile video streaming, the authors in [5] derived an optimal solution for network-assisted adaptation in multi-access edge computing environments that can assist the dynamic adaptive streaming over hypertext transfer protocol system in large-scale scenarios. The solution includes a client-to-server mapping strategy and the joint weighted maximization of QoE and fairness, while utilizing the system resource.

By bringing computing capabilities to the network edge, an effective video transcoding and delivery approach based on mobile edge computing and blockchain was proposed in [6, 7]. Incentive mechanism-aware offloading schedule, block size adaptation scheme, and blockchain technique were considered in order to maximize the average transcoding profit for the transcoders and enable low-latency MAS. Especially in [7], small-cell base stations (SBS) equipped with mobile edge computing servers can assist in providing MAS to MUs in a distributed, incentivized, and secure manner. Thus, the content providers are able to autonomously adjust their strategies according to video content market statistics by allocating their resources and determining content prices associated with quality levels in a three-stage Stackelberg game.

1.3.2 Multicasting Assisted Multimedia Communications

Multicasting is the most efficient technique for group-oriented services to serve MUs within a particular area who have the same content interest. Various solutions related to multicasting techniques have been studied in [8–13]. A joint solution of subgrouping and non-orthogonal multiple access (NOMA) for multicasting MAS in 5G networks was studied in [8] to maximize the sum rate received by all the MUs, while saving the spectrum and bandwidth system resources. In [9, 10], because the evolved multimedia broadcast multicast services cannot support crowdsourced live MAS, the authors designed a new service-less video multicast at the edge in 5G networks to reduce the traffic burden on different levels. A virtual network function collects the potential video requests and then initiates the ad hoc multicast MAS on the fly. The proposed solutions can improve the QoE, provide a cost-savings solution in fetching the data from the content delivery networks (CDN), and enhance spectrum efficiency on transit, backhaul, and radio access links.

To improve the performance of the multicasting technique, multicasting was combined with ABR [11] and power allocation [12]. Different transmission powers were allocated to different bitrates (layer streams) to minimize the total target rate while ensuring the proportional rate constraints from the high-priority base layer stream to low-priority additional layer streams [12]. This solution can also provide a robust delivery of standard-quality video to all MUs in NOMA-based multilayer multicast systems. In [13], the transmission powers were allocated to multiple video layers in accordance with their channel conditions. As a result, the proposed method can outperform the conventional NOMA-based algorithms in terms of QoE under various multicast scenarios.

1.3.3 D2D Multimedia Communications

D2D communications have been playing an important role as the communication backbone in edge techniques for MAS, via direct links over one hop [14–18] or indirect links over multiple hops [19, 20]. More importantly, D2D communications can enable other emerging solutions, i.e., resource allocation, scheduling, mode selection, ABR, and social relationship, to provide MUs with MAS of high QoE. The D2D communications techniques associated with other emerging solutions are detailed below.

In [14], the authors proposed an edge framework for video streaming in D2D networks in which multiple D2D pairs for MAS are established by applying spectrum allocation, mode selection, and power control. Scalable video coding is utilized to transmit high-quality videos over the good D2D links while ensuring basic-quality videos over the bad D2D links. This results in a better performance in terms of average playback quality for all the MUs. To improve the work in [14], D2D pair association solution was considered in [15]. The association solution pairs a requesting MU with a helping MU, to gain the highest playback quality while consuming less energy for D2D communications.

A cross-layer design for D2D video communications in cooperative cognitive radio networks was proposed in [16], where adaptive modulation and coding at the physical layer, automatic repeat request at the link layer, and scalable video coding at the application layer are synthesized into a cognitive radio network structure to simultaneously maximize the QoE and minimize the power consumption for D2D pairs. In [17], the authors studied a social-aware method based on mobility similarity measurement for D2D video delivery, showing that when the social relationship between the two MUs of a D2D pair is considered, the encounter durations of D2D pairs and the video lengths can be matched well to establish the best video streaming sessions over D2D communications. Furthermore, multi-access edge computing technology can be applied to MAS over D2D communications so that the knowledge of mobile devices in proximity can be maintained [18]. As a result, the discovery procedure for D2D video streaming session is simplified to reduce the network latency.

D2D communications can be expanded to multiple hops so that more resources around the requesting MUs are discovered and reused for MAS [19–21]. This benefits from short transmission distance to spread the low-power consumption in D2D networks and thus causes low-interference impact on the MUs who share their spectrum resources for D2D communications. To establish a multi-hop path, routing is preliminarily required, and then resource allocation can be applied for video streaming sessions as studied in [19, 20]. In [21], multi-hop D2D networks are assisted by caching to further exploit the videos and available storage resources located longer than one hop far away from the requesting MU. By cache-aided multi-hop D2D networks, the requesting MUs have an alternate way to entertain the MAS besides being served by direct D2D communications or macro base stations (MBS).

1.3.4 Caching Assisted Multimedia Communications

As one of the most important edge techniques for MAS, caching exploits the available edge storage resource of MBS, SBS, roadside units, UAV, and mobile devices to bring multimedia contents closer to MUs. Consequently, it can mitigate the congestion at the backhaul links of SBS and MBS, provide high QoE, and conserve the system resources. There are two typical caching techniques: deterministic caching [22] and probabilistic caching [23]. Deterministic caching deploys the contents in stable edge placements such as MBS, SBS, and/or roadside units. Meanwhile, probabilistic caching deploys in dynamic placements, e.g., D2D, UAV, vehicular ad hoc networks (VANETs), and even SBS as long as MUs can join/leave the caching networks randomly.

While deterministic caching techniques have been well investigated in the literature, probabilistic caching techniques for content delivery networks have recently gained popularity among studies in D2D networks, UAV, and vehicular networks. The probabilistic caching techniques aim to find the optimal caching probability for each content [23, 24] or are combined with other techniques (spectrum partitioning, channel access, cooperative transmission, clustering, energy harvesting, and transmission mode) to enhance the performance of D2D caching networks [25–31]. However, most studies considered QoS-based performance metrics such as cache hit probability, delay, offloading gain, cache-added throughput, and cache-aided successful transmission probability for common contents, rather than QoE for MAS. In addition, the following limitations remain in caching network design: (i) contents are assumed to be in equal unit size, (ii) matching resources between the capacity (represented by the throughput) of the system and playback resolution requirement of mobile devices is overlooked, and thus rate–distortion model-based ABR for caching is not applied, (iii) power allocation is not considered, and (iv) MBS is not considered for assisting in cooperative transmission. As a result, conventional methods do not take full advantage of system resources, characteristics of videos and mobile devices, and additional techniques, to provide MAS of highest QoE.

Related to probabilistic caching at the SBS, the dynamic on-off stage of SBS for caching the popular contents was considered in [32, 33]. The caching probabilities are optimized to maximize the successful download probability and reduce the duplicated contents, which cause wasted transmission power. The authors in [33] extended the study in [32] by further considering probabilistic caching for ultra-dense SBS networks with terrestrial and aerial users. Various caching performance scenarios were analyzed to provide valuable insights into the SBS caching networks. The probabilistic caching deployed in VANET was studied in [34], considering MUs as vehicular users. The contents are distributed based on the cooperation of vehicular users and roadside units. As such, the vehicular users are able to request the contents from others or from fixed roadside units with a high hit rate and less caching overhead.

Another efficient probabilistic caching in VANET has been proposed in [35], where the caching strategy is adjusted in accordance with vehicular users' demand, vehicle importance degree, and relative movement of the transmitter and receiver. Although complicated, it can provide the vehicular users with a high hit rate and low delay but fewer replicas in dense VANET. Moreover, cooperative caching in static and dynamic vehicles was shown to yield a significant improvement in terms of delay, packet delivery ratio, and cache hit rate compared to other noncooperative solutions [36].

Probabilistic caching becomes more powerful with "caching to the sky" done by UAV [37–39]. The UAV can flexibly focus on a particular area for caching and exploit the benefit of line-of-sight transmission to provide both the stable and dynamic MU with MAS of high QoE. In [37], the authors introduced a joint static and dynamic probabilistic caching at UAV to assist D2D caching networks, where static caching is to maximize the cache hit rate and dynamic caching is to further fulfill the coverage for all MUs while minimizing the number of path points for shortest trajectory. The authors in [38] further considered the movement of MUs in UAV-D2D-enabled caching networks. K-means clustering is utilized to obtain the partition of MUs and a mobility-aware probabilistic caching algorithm is designed to maximize the total cache-aided throughput. The so-called "caching to the sky" can work efficiently with VANET [39]. Due to the complicated dynamic environment of UAV-VANET, the authors proposed a comprehensive framework for modeling and analyzing the cooperative probabilistic UAV-VANET caching solution based on stochastic geometry but in an approximate approach. The study provides useful insights into the design of cooperative probabilistic caching in UAV-VANET.

1.4 Multimedia Communications in 6G Networks: Challenges and Research Directions

1.4.1 Challenges

As discussed earlier, the primary concern in 6G multimedia communications is the demand from a massive number of MUs requesting the ever-high data rate MAS. In what follows, we discuss the main challenges in edge computing, multicasting, D2D, and caching-assisted multimedia communications.

In edge computing, while utilizing computing capabilities at the network edge can improve the performance of MAS, most studies have not exploited edge resources of dynamic platforms such as UAV and VANET [4–7]. These resources, when used, can catch up with the varying characteristics of MUs (i.e., mobility, density, common interest, and social relationship) to provide MAS more efficiently, while acting as a relay network to expand network coverage, and thus maximize the number of MUs with better communications thanks to a higher opportunity to gain line-of-sight propagation assisted by UAV.

Multicasting technique is promising in terms of improving QoE and saving system resources for local MAS in which the MUs have the same content interests

[8–13]. It becomes more efficient when combined with other techniques such as subgrouping, NOMA, power allocation, and rate adaptation. However, multicasting techniques cannot guarantee fairness among MUs due to various wireless channel conditions and diverse social relationships. In addition, multicasting sources at the network edge (from UAV and VANET) have not been exploited to further satisfy dynamic MUs in dense networks.

While D2D communications can enable various emerging solutions to provide the MUs with better MAS, it often operates in a stand-alone network and does not introduce any disruptive edge architecture for MAS. D2D communications networks should be integrated with other edge networks such as UAV and vehicular networks so that the entire edge architecture benefits from many techniques and solutions to provide a better MAS experience.

Regarding caching, the aforementioned probabilistic caching techniques in existing studies have focused on general content rather than MAS. Meanwhile, multimedia aspects and MU characteristics play an important role in MAS system design. For example, the playback resolution of the mobile devices must be considered to exploit the system bandwidth and rate adaptation for satisfying the MUs, especially in dynamic environments. Moreover, QoE is preferred to QoS as the performance metrics for MAS. QoE is represented by a set of metrics happening in order from the time when an MU requests a multimedia content to the time when it leaves or stops watching. This set includes latency, hit rate, continuous playback, high playback quality, and low quality fluctuation. Without these considerations, the probabilistic caching technique is not sufficient to serve the MUs high QoE while utilizing the system resources.

Finally, the aforementioned solutions cannot be efficiently applied to MAS in various types of dense networks for the following reasons: firstly, these solutions are separate from each other; their advantages cannot be combined to improve network performance, flexibility, and efficiency. Secondly, many MU-related aspects have not been adequately addressed, including complex access behaviors, social relationships, mobility patterns, and the diverse capacities of mobile devices. Thirdly, the QoE in MAS is highly sensitive to various factors such as wireless transmission environment, MU characteristics, system resources, and multimedia features and codec techniques. Importantly, high QoE of MUs is presented by five performance metrics: low latency, high hit rate to access the videos, punctual arrival for continuous playback, high visual playback quality measured in peak signal-to-noise ratio, and low quality fluctuation for smooth playback. Last but not least, system resources (spatial, time, storage, energy, bandwidth, and spectrum) have not been fully exploited to improve the performance of multimedia communications.

1.4.2 Research Directions

To meet the requirements of 6G multimedia communications, developing high-speed backhaul links with new architectures and technologies for MUs has been considered, but it is costly. Alternatively, these requirements can be efficiently addressed through infrastructure transformation that places a higher priority on software-defined everything. This infrastructure transformation primarily relies on edge techniques combined with intent-driven and context-aware solutions for MAS, i.e., bringing system resources closer to high-demand areas and leveraging the nearby resources of dense MUs to achieve better QoE. Internet service and content providers can further satisfy the MUs' demand without changing network architectures, and thus at a lower cost. The most critical issue is the lack of a comprehensive system design framework for edge techniques and assisted solutions, as illustrated in Figure 1.3.

First, all aspects of MUs (mobility, behavior, social relationships, and playback capacity/resolution), the characteristics of multimedia content (size, popularity, codec, packetization scheme, and rate–distortion model), and the features of devices and networks (integration, cooperation, and resources) are examined. Next, these aspects, characteristics, and features are analyzed and modeled for the design of multimedia communication systems. A multimedia communication system design framework takes into account the edge techniques

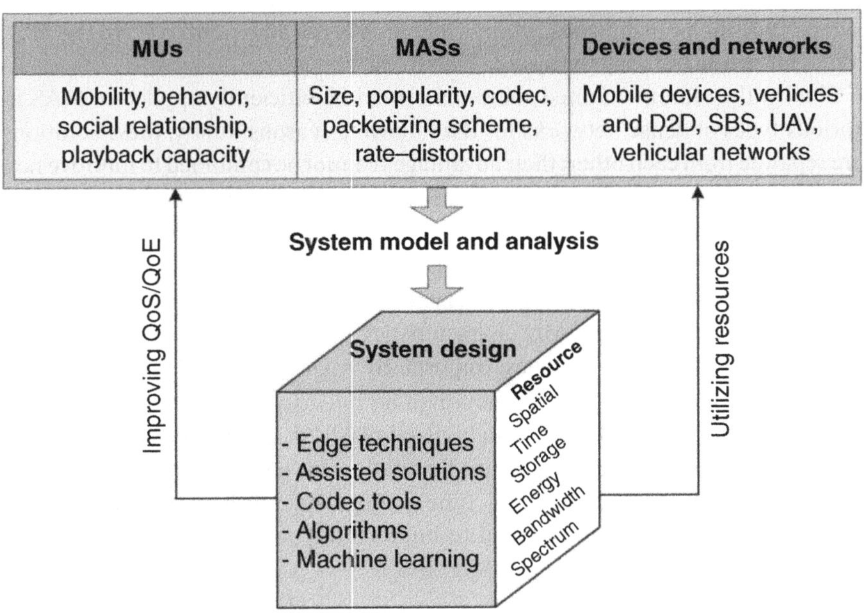

Figure 1.3 A framework for 6G multimedia communication systems.

(edge computing, multicasting, D2D communications, and caching), assisted solutions (ABR, incentive mechanism, cross-layer design, user association, resource allocation, clustering, subgrouping, NOMA, mode selection, channel access, cooperative transmission, energy harvesting, routing, and intelligent reflecting surfaces), video codec tools (HEVC/H.265), algorithms, and machine learning. The framework also utilizes the system resources (spatial, time, storage, energy, bandwidth, and spectrum) across various types of devices and networks for resource efficiency.

The optimization problems of MAS are formulated in the form of maximizing QoS/QoE functions with respect to a number of variables (e.g., caching placement, caching probability, cache density, ABR and packetizing, user-caching placement association, power allocation, spectrum sharing, cluster radius, and 3D trajectory) under constrained system resources. The optimal video packetizing scheme obtained from solving the optimization problems is actually the optimal ABR to partition each video or video segment into different layers. These layers are then packetized into multiple packets (descriptions) for delivering using layered multiple descriptions coding with embedded forward error correction. This ensures that the MAS is protected against error-prone transmission and adaptively sent over the diverse transceiver bandwidth and playback capacities of mobile devices and varying characteristics of wireless networks.

1.5 Conclusion

We have presented an overview of multimedia communications, advanced applications and services, and MAS in 6G networks. The emerging techniques for 6G multimedia communications were reviewed, analyzed, and discussed. Our discussion revealed the challenges and research directions of 6G multimedia communications. This chapter has provided the concepts, models, techniques, solutions, and background of multimedia communications that are further studied in the subsequent chapters.

References

1 V. Lionel Sujay. Number of IoT connections worldwide 2022–2033. *Statista Research Department*. https://www.statista.com. 2024.

2 N.-S. Vo, T.-M. Phan, M.-P. Bui, X.-K. Dang, V. T. Nguyen, and C. Yin. Social-aware spectrum sharing and caching helper selection strategy optimized multicast video streaming in dense D2D 5G networks. *IEEE Syst. J.*, 15(3):3480–3491. 2021.

3 T. T. Le, D. V. Nguyen, and E.-S. Ryu. Computing offloading over mmWave for mobile VR: Make 360 video streaming alive. *IEEE Access*, 6:66576–66589. 2018.

4 L. Li, R. Hou, R. Li, H. Li, M. Pan, and Z. Han. Delay-aware adaptive wireless video streaming in edge computing assisted ultra-dense networks. In *Proc. IEEE Global Commun. Conf.*, pages 1–6, Waikoloa, HI, Dec. 2019.

5 A. Mehrabi, M. Siekkinen, and A. Ylä-Jääski. Edge computing assisted adaptive mobile video streaming. *IEEE Trans. Mob. Comput.*, 18(4):787–800. 2018.

6 M. Liu, F. R. Yu, Y. Teng, V. C. M. Leung, and M. Song. Distributed resource allocation in blockchain-based video streaming systems with mobile edge computing. *IEEE Trans. Wirel. Commun.*, 18(1):695–708. 2018.

7 Y. Liu, F. R. Yu, X. Li, H. Ji, and V. C. M. Leung. Decentralized resource allocation for video transcoding and delivery in blockchain-based system with mobile edge computing. *IEEE Trans. Veh. Technol.*, 68(11):11169–11185. 2019.

8 J. Montalban, P. Scopellitti, M. Fadda, E. Iradier, C. Desogus, P. Angueira, M. Murroni, et al. Multimedia multicast services in 5G networks: subgrouping and non-orthogonal multiple access techniques. *IEEE Commun. Mag.*, 56(3): 91–95. 2018.

9 K. Zahoor, K. Bilal, A. Erbad, and A. Mohamed. Service-less video multicast in 5G: Enablers and challenges. *IEEE Netw.*, 34(3):270–276. 2020.

10 K. Zahoor, K. Bilal, A. Erbad, A. Mohamed, and M. Guizani. Multicast at edge: An edge network architecture for service-less crowdsourced live video multicast. *IEEE Access*, 9:59508–59526. 2021.

11 X. Tan, L. Xu, Q. Zheng, S. Li, and B. Liu. QoE-driven DASH multicast scheme for 5G mobile edge network. *J. Commun. Inform. Netw.*, 6(2):153–165. 2021.

12 M. N. Dani, D. K. C. So, J. Tang, and Z. Ding. Resource allocation for layered multicast video streaming in NOMA systems. *IEEE Trans. Veh. Technol.*, 71(11):11379–11394. 2022.

13 M. Zhang, H. Lu, F. Wu, and C. W. Chen. NOMA-based scalable video multicast in mobile networks with statistical channels. *IEEE Trans. Mob. Comput.*, 20(6):2238–2253. 2021.

14 I. Ahmed, M. H. Ismail, and M. S. Hassan. A QoE-aware mode selection framework for video streaming in D2D networks. *IEEE Access*, 8:169272–169285. 2020.

15 S. Ullah, K. Kim, A. Manzoor, L. U. Khan, S. M. A. Kazmi, and C. S. Hong. Quality adaptation and resource allocation for scalable video in D2D communication networks. *IEEE Access*, 8:48060–48073. 2020.

16 S. Chatterjee and S. De. QoE-aware cross-layer adaptation for D2D video communication in cooperative cognitive radio networks. *IEEE Syst. J.*, 16(2):2078–2089. 2022.

17 R. Zhang, S. Jia, Y. Ma, and C. Xu. Social-aware D2D video delivery method based on mobility similarity measurement in 5G ultra-dense network. *IEEE Access*, 8:52413–52427. 2020.

18 S.-R. Yang, C.-J. Shih, and P. Lin. Multi-access edge computing-assisted D2D streaming for proximity-based social networking. In *Proc. IEEE Global Commun. Conf.*, pages 1–6, Waikoloa, HI, Dec. 2019.

19 S. Alwan, I. Fajjari, and N. Aitsaadi. Joint routing and wireless resource allocation in multihop LTE-D2D communications. In *Proc. IEEE Local Computer Netw. Conf.*, pages 1–8, Chicago, IL, Oct. 2018.

20 J. Li, T. Li, F. Li, T. Hong, and X. Wang. Multi-hop D2D assisted real-time video streaming transmission system in infrastructure-less networks. In *Proc. Inter. Computer Commun. and Netw. Conf.*, pages 1–6, Valencia, Spain, July 2019.

21 M.-C. Lee, M. Ji, and A. F. Molisch. Optimal throughput-outage analysis of cache-aided wireless multi-hop D2D networks. *IEEE Trans. Commun.*, 69(4):2489–2504. 2021.

22 N.-S. Vo, M.-P. Bui, P. Q. Truong, C. Yin, and A. Masaracchia. Multi-tier caching and resource sharing for video streaming in 5G ultra-dense networks. *IEEE Commun. Lett.*, 24(7):1500–1504. 2020.

23 Z. Chen, N. Pappas, and M. Kountouris. Probabilistic caching in wireless D2D networks: Cache hit optimal versus throughput optimal. *IEEE Commun. Lett.*, 21(3):584–587. 2017.

24 J. Ma, L. Liu, B. Shang, and P. Fan. Cache-aided cooperative device-to-device (D2D) networks: A stochastic geometry view. *IEEE Trans. Commun.*, 67(11):7444–7455. 2019.

25 R. Amer, H. ElSawy, J. Kibiłda, M. M. Butt, and N. Marchetti. Cooperative transmission and probabilistic caching for clustered D2D networks. In *Proc. IEEE Wireless Commun. and Networking Conf.*, pages 1–6, Marrakesh, Morocco, Apr. 2019.

26 R. Amer, H. Elsawy, M. M. Butt, E. A. Jorswieck, M. Bennis, and N. Marchetti. Optimized caching and spectrum partitioning for D2D enabled cellular systems with clustered devices. *IEEE Trans. Commun.*, 68(7):4358–4374. 2020.

27 R. Amer et al. Optimizing joint probabilistic caching and channel access for clustered D2D networks. *Journal of Commun. and Netw.*, 23(6):433–441. 2021.

28 Y. Meng, Z. Zhang, and Y. Huang. Cache- and energy harvesting-enabled D2D cellular network: Modeling, analysis and optimization. *IEEE Trans. Green Commun. and Netw.*, 5(2):703–713. 2021.

29 M. Naslcheraghi, C. Wette, J.-F. Frigon, and B. Sansò. Probabilistic analysis of operating modes in cache-enabled full-duplex D2D networks. *IEEE Trans. Veh. Technol.*, 71(6):6624–6635. 2022.

30 X. Song, Y. Geng, X. Meng, J. Liu, W. Lei, and Y. Wen. Cache-enabled device to device networks with contention-based multimedia delivery. *IEEE Access*, 5:3228–3239. 2017.

31 J. Ma, L. Liu, H. Song, R. Shafin, B. Shang, and P. Fan. Scalable video transmission in cache-aided device-to-device networks. *IEEE Trans. Wirel. Commun.*, 19(6):4247–4261. 2020.

32 Y. Chen, M. Ding, J. Li, Z. Lin, G. Mao, and L. Hanzo. Probabilistic small-cell caching: Performance analysis and optimization. *IEEE Trans. Veh. Technol.*, 66(5):4341–4354. 2017.

33 F. Song, J. Li, M. Ding, L. Shi, F. Shu, M. Tao, W. Chen, et al. Probabilistic caching for small-cell networks with terrestrial and aerial users. *IEEE Trans. Veh. Technol.*, 68(9):9162–9177. 2019.

34 Z. Su, Y. Hui, Q. Xu, T. Yang, J. Liu, and Y. Jia. An edge caching scheme to distribute content in vehicular networks. *IEEE Trans. Veh. Technol.*, 67(6):5346–5356. 2018.

35 G. Deng, L. Wang, F. Li, and R. Li. Distributed probabilistic caching strategy in VANETs through named data networking. In *Proc. IEEE Conf. Computer Commun. Workshops*, pages 314–319, San Francisco, CA, Apr. 2016.

36 S. A. Elsayed, S. Abdelhamid, and H. S. Hassanein. Probabilistic cooperative caching in VANETs for social networking. In *Proc. IEEE Global Commun. Conf.*, pages 1–7, Abu Dhabi, United Arab Emirates, Dec. 2018.

37 J. Ji, K. Zhu, D. Niyato, and R. Wang. Probabilistic cache placement in UAV-assisted networks with D2D connections: Performance analysis and trajectory optimization. *IEEE Trans. Commun.*, 68(10):6331–6345. 2020.

38 Y. J. Chen, K. M. Liao, M. L. Ku, and F. P. Tso. Mobility-aware probabilistic caching in UAV-assisted wireless D2D networks. In *Proc. IEEE Global Commun. Conf.*, pages 1–6, Waikoloa, HI, Dec. 2019.

39 S. Zhao, Y. Zhang, Y. Zhu, Z. Zhao, and Y. Liu. Optimal probabilistic collaborative caching in UAV-assisted vehicular networks. In *Proc. Inter. Intell. Transport. Syst. Conf.*, pages 377–382, Macau, China, Nov. 2022.

2

Encoding Rate Selection and Probabilistic Caching for Video Offloading in Dense D2D Networks

2.1 Introduction

In the 6G era, numerous MUs in extremely dense networks are more inclined to use the advanced applications and services, e.g., social media, e-health, e-education, intelligent transportation, public safety, city government, smart business, gaming, and entertainment.[1] Although 6G networks have been stimulated to prepare for what 5G could not fulfill [2], the huge amount and high data rate of advanced applications and services, especially multimedia applications and services (MAS), might outpace the capacities of Internet service and content providers. In this context, investment in architectural modifications and technology upgrades is a necessary prerequisite, but obviously costly. Meanwhile, software-defined techniques can leverage 6G networks for multimedia communications through flexible, rapid, and low-cost deployment, not only by utilizing advanced architectures and technologies but also by efficiently allocating system resources.

In 6G MAS, software-defined techniques primarily rely on intent-driven and context-aware data methodologies to enhance user-perceived quality. For example, popular content can be cached closer to high-demand areas [3, 4], MUs with similar content interests can be clustered into groups for multicasting [5, 6], and multimedia content can be encoded at multiple bitrates to match the diverse capabilities of mobile devices and wireless channels [1]. In dense D2D networks, the caching technique plays an important role in providing local 6G MAS thanks to the use of available resources of nearby mobile devices with the assistance of D2D communications [7–9].

Various caching strategies have been studied, mostly focusing on (i) where and how to implement caching, and (ii) the capability to deliver content after caching. To deal with the former consideration, deterministic or probabilistic caching is used. Deterministic caching deals with determining cache placements for each

1 This chapter has been published partly in [1].

content in stable locations, such as terrestrial macro base stations (MBS) and small-cell base stations [3]. Although deterministic caching is difficult to implement in large-scale networks, it is well suited for serving particular MU groups. Meanwhile, probabilistic caching aims to determine the caching probability of each content in dynamic and mobile placements within D2D networks. Its main advantage is the ease of implementation, allowing the MUs to join or leave caching networks randomly [10]. For the latter consideration, additional solutions are integrated into caching techniques to properly configure the system parameters, such as transmission power, channel allocation, and transmission mode selection, for delivering the cached content to the MUs. Each caching technique has its own advantages and limitations. This chapter is dedicated to studying the probabilistic caching in D2D networks.

Probabilistic caching has been considered as an emerging technique for content offloading in D2D networks [10–17]. Particularly, the authors in [10, 11] studied how to enhance the QoS-based performance metrics, e.g., cache hit probability, cache-added throughput (data rate), and cache-aided successful probability of transmission (content delivery), by finding the caching probability of each content. Integrating spectrum partitioning, channel access, energy harvesting, and transmission mode selection solutions with probabilistic caching can improve system performance [12–15].

However, most studies on probabilistic caching have not addressed existing issues in caching optimization design for MAS. Firstly, all the contents were assumed to have the same unit size, which is not practical. The caching optimization performance is degraded as the storage resource cannot be allocated well with respect to the skewed popularity pattern of contents. Secondly, mobile devices have different playback resolution capabilities. It is challenging to satisfy the MUs with local video applications and services (VAS) by utilizing system throughput resources while ignoring statistical knowledge of the diverse playback resolution requirements of the mobile devices. Thirdly, the objective functions are derived from QoS-based performance metrics instead of visual user-perceived quality-based ones. Finally, adaptive bitrate (ABR) solution has not been applied to encoding the videos into multiple rates with different quality versions. As a result, they cannot simultaneously satisfy the diverse playback resolution requirements of the mobile devices and the varying characteristics of the wireless environment, nor can they fully utilize system throughput and storage resources for caching and delivery.

In this chapter, we address the above problems by proposing a joint encoding rate selection and probabilistic caching (RSC) optimization solution for video offloading in dense D2D networks. We consider the practical aspects of videos (multiple encoding rates with different sizes) and MUs (playback resolution requirement and performance metric based on user-perceived quality). By using rate–distortion model (RDM) for video encoding, homogeneous Poisson Point

Process model for MUs' distribution, and D2D underlay transmission mode for video offloading, the RSC problem is formulated with respect to real and binary optimization variables for the optimal caching probability and encoding rate selection, respectively. Genetic algorithms (GAs) are feasibly applied to solve the RSC optimization problem. Simulation results are shown to demonstrate the benefits of the RSC method in terms of playback quality and system resource conservation. The findings are also analyzed to provide useful insights into caching optimization design for VAS.

The rest of this chapter is organized as follows. In Section 2.2, we introduce the RSC system model for video offloading. The formulation of the RSC model is analyzed in Section 2.3. We present the RSC optimization problem and propose the solution in Section 2.4. In Section 2.5, the performance of the proposed RSC method is evaluated. Finally, Section 2.6 concludes the chapter.

2.2 RSC System for Video Offloading

2.2.1 RSC Components

The RSC system for video offloading in dense D2D networks is illustrated in Figure 2.1. The system consists of one MBS covering a circular radius R_{MBS}, I videos, and a large number of MUs. Video i is encoded into J_i quality versions $v_{i,j}$ ($i = 1, 2, ..., I$; $j = 1, 2, ..., J_i$), each with encoding rate $R_{i,j}$ Kbps and size $S_{i,j}$ Kbit. The MUs in the vicinity of D2D communications having the same video interest are grouped into a cluster with radius R_{D2D}. There are two types of MUs: helping MUs (HUs) and requesting MUs (RUs). The HUs contribute their storage resources to cache the videos (video helping MUs—VHs) and share their

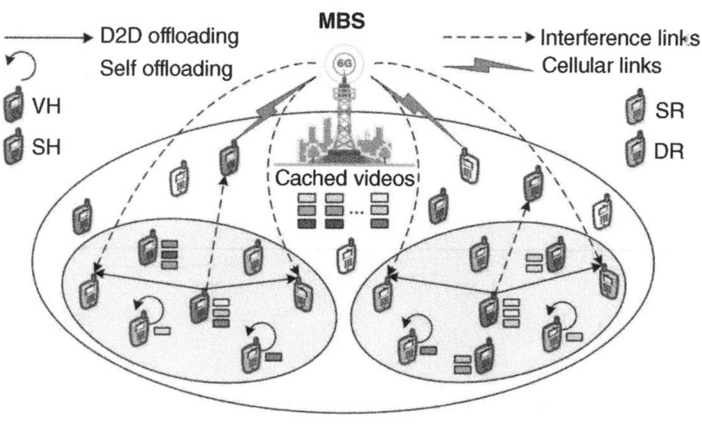

Figure 2.1 RSC system for video offloading in dense D2D networks.

spectrum for offloading the videos (spectrum helping MUs—SHs). The RUs, who are interested in watching the videos, include video self-requesters (SRs) and video D2D-requesters (DRs). For each requested video, the SRs have already cached it, whereas the DRs have not.

2.2.2 Preliminary RSC Models

2.2.2.1 Video Popularity Model

In VAS, each video has its own popularity due to the interest behavior of MUs. Analyzing the interest behavior of the MUs enables us to acquire the popularity pattern of videos, thus enhancing the performance of the RSC method, i.e., achieving better playback quality while maintaining higher resource allocation efficiency. Following the Zipf-like distribution [9], the popularity (or access rate) of video i is given by

$$r_i = \frac{i^{-\alpha}}{\sum_{l=1}^{I} l^{-\alpha}}, \tag{2.1}$$

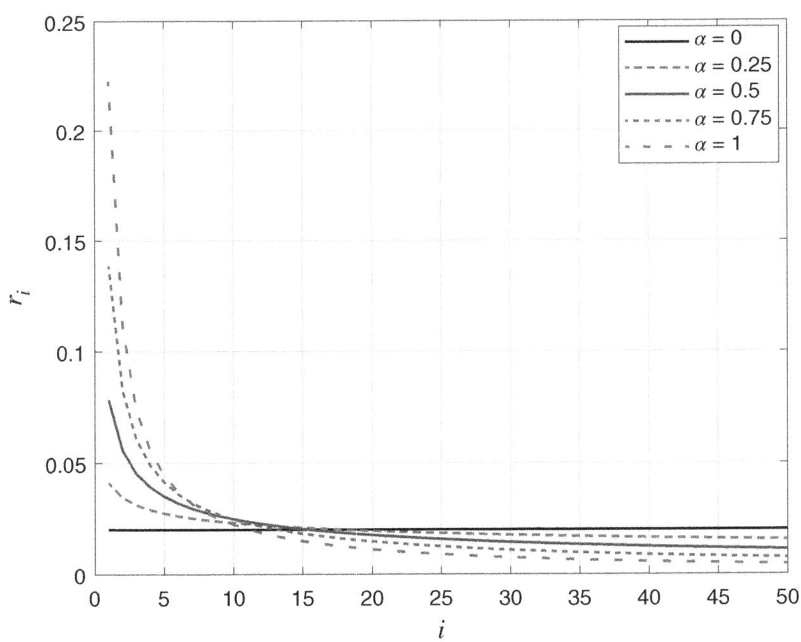

Figure 2.2 Video popularity patterns with different skewed coefficients α.

where $\alpha \geq 0$ is the skewed popularity coefficient among I videos. As shown in Figure 2.2, changing α yields different video popularity patterns reflecting the interest behavior of MUs. Especially, if $\alpha = 0$, all videos have the same popularity of $1/I$.

2.2.2.2 Rate-distortion Model

Video i of encoding rate $R_{i,j}$ is decoded for playing back at a reconstructed distortion of $D_{i,j}$ measured in mean squared error (MSE). The relationship between $D_{i,j}$ and $R_{i,j}$ can be modeled as a decaying exponential function [3]

$$D_{i,j} = \gamma_i (R_{i,j})^{\beta_i}, \tag{2.2}$$

where γ_i and β_i are found to match the experimental RDM of $v_{i,j}$. In Figure 2.3, we analyze the two well-known videos named Foreman and Mobile by using HM reference software version 12.0 [18] to obtain their experimental RDMs. Then, we find γ_i and β_i to generate (2.2) that matches the experimental RDMs perfectly. The decaying exponential function is more conventionally considered in the RSC optimization problem, which is solved to select $v_{i,j}$ for caching and offloading.

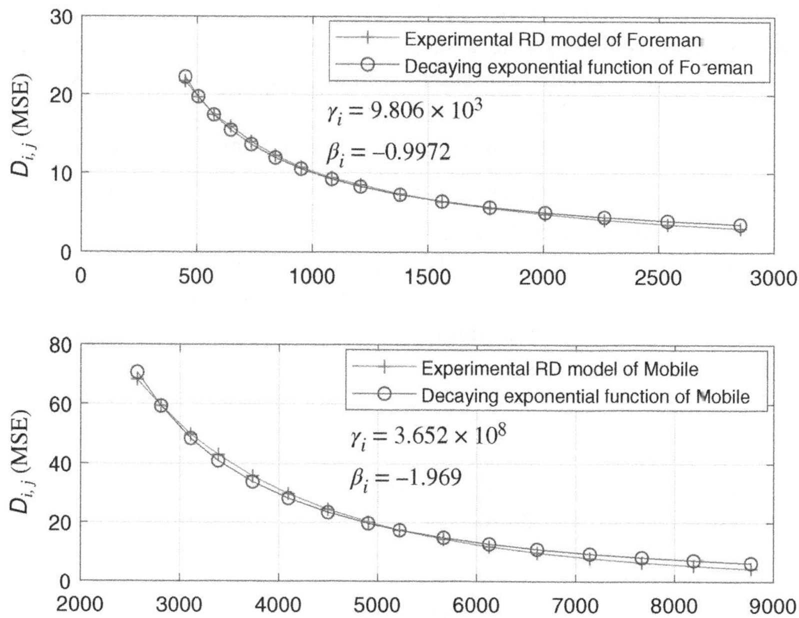

Figure 2.3 A perfect match between decaying exponential function and experimental RDM.

2.2.2.3 Homogeneous Poisson Point Process Model

The distribution of MUs is modeled by a homogeneous Poisson Point Process Φ_{MU} with density λ_{MU}. Each MU can act as an HU with probability ρ or an RU with probability $(1-\rho)$. An HU refers to either (i) a VH that caches $v_{i,j}$ with probability p_i or (ii) an SH that does not cache $v_{i,j}$, with probability $(1-p_i)$, but shares its downlink spectrum resources for offloading $v_{i,j}$ via D2D communications. Similarly, an RU refers to either an SR with probability p_i, since the SR has cached $v_{i,j}$, or a DR with probability $(1-p_i)$, since the DR has not yet cached $v_{i,j}$. Thus, the distributions of VH, SH, SR, and DR follow the homogeneous Poisson Point Process Φ_i^{VH}, Φ_i^{SH}, Φ_i^{SR}, and Φ_i^{DR} with densities λ_i^{VH}, λ_i^{SH}, λ_i^{SR}, and λ_i^{DR}, respectively, as shown in Table 2.1.

2.2.3 RSC Operation Description

The MBS is in charge of implementing the RSC system. It first collects the necessary information about the MUs, the videos, and the wireless channels to formulate the RSC optimization problem. The objective is to maximize the playback quality by finding the optimal caching probability p_i and the optimal encoding rate selection index $w_{i,j} \in \{0, 1\}$ under the constraints on the throughput of system and the caching storage of mobile devices. Here, $w_{i,j} = 1$ indicates that $v_{i,j}$ is selected for caching with probability p_i; otherwise, $w_{i,j} = 0$. Then, the MBS groups the MUs into different clusters separated by the cluster radius R_{D2D}. So far, the system is ready to provide the RUs with local VAS. Particularly, when there is a request for a video from the RUs in a cluster, the system works in two modes as follows:

- Self-offloading (SOF) mode: The SRs have cached the requested video earlier, and thus they are served by themselves.
- D2D-offloading (DOF) mode: The DRs have not cached the requested video, but at least one of the VHs within the radius R_{D2D} have cached it. The DRs are served by the VHs over D2D communications reusing the downlink spectrum resource shared by the SHs.

Table 2.1 MUs' densities.

Densities	Expressions	Descriptions
λ_i^{VH}	$\rho p_i \lambda_{\text{MU}}$	Video helping MU
λ_i^{SH}	$\rho(1-p_i)\lambda_{\text{MU}}$	Spectrum helping MU
λ_i^{SR}	$(1-\rho)p_i\lambda_{\text{MU}}$	Self-requesting MU
λ_i^{DR}	$(1-\rho)(1-p_i)\lambda_{\text{MU}}$	D2D-requesting MU

2.3 Analysis and Formulation of RSC Model for Video Offloading

In this section, we formulate the RSC model for video offloading in dense D2D networks, including hit probabilities, achievable rate probability, average playback quality, and throughput and storage resource consumption.

2.3.1 Hit Probabilities

Since the MUs can act as HUs (VHs for caching and SHs for spectrum sharing) or RUs (SRs for SOF mode and DRs for DOF mode), the helping probabilities include cache hit probability and spectrum sharing hit probability, while the requesting probabilities include self-requesting hit probability and D2D-requesting hit probability. The self-requesting hit probability is defined as p_i. The other hit probabilities are computed by following the homogeneous Poisson Point Process Φ_{MU} with density λ_{MU} in which the probability of having N MUs within a circular radius of R_{X} is given by [10].

$$\Pr(N, R_{\mathrm{X}}, \lambda_{\mathrm{MU}}) = \frac{(\pi R_{\mathrm{X}}^2 \lambda_{\mathrm{MU}})^N}{N!} e^{-\pi R_{\mathrm{X}}^2 \lambda_{\mathrm{MU}}}, \mathrm{X} \in \{\mathrm{MBS}, \mathrm{D2D}\}. \tag{2.3}$$

Based on (2.3), within R_{D2D}, the cache hit probability ($\Pr_i^{\mathrm{VH}}$) is the probability of at least one MU caching video i and the D2D-requesting hit probability ($\Pr_i^{\mathrm{DR}}$) is the probability of at least one MU requesting video i, given by

$$\Pr_i^{\mathrm{Y}} = 1 - \Pr(N = 0, R_{\mathrm{D2D}}, \lambda_i^{\mathrm{Y}}) = 1 - e^{-\pi R_{\mathrm{D2D}}^2 \lambda_i^{\mathrm{Y}}}, \mathrm{Y} \in \{\mathrm{VH}, \mathrm{DR}\}. \tag{2.4}$$

Similarly, within R_{MBS}, the spectrum sharing hit probability is

$$\Pr_i^{\mathrm{SH}} = 1 - e^{-\pi R_{\mathrm{MBS}}^2 \lambda_i^{\mathrm{SH}}}. \tag{2.5}$$

2.3.2 Achievable Rate Probability

Achievable rate probability is the probability that the channel capacity is greater than or equal to the necessary rate required to offload $v_{i,j}$. In SOF mode, the achievable rate is $\Pr_{i,j}^{\mathrm{SA}} = 1$. In DOF mode, we need to derive the achievable rate probability $\Pr_{i,j}^{\mathrm{DA}}$ at DR m, $m \in \Phi_i^{\mathrm{DR}}$. To do so, we compute the channel capacity from VH n, $n \in \Phi_i^{\mathrm{HC}}$, to DR m, expressed as

$$C_{n,m}^{(i)} = W \log_2 \left(1 + \frac{P_{\mathrm{D2D}} g_{n,m}}{\sigma_0 + P_{\mathrm{MBS}} g_{\mathrm{MBS},m} + \sigma_i}\right), \tag{2.6}$$

where W is the system bandwidth; P_{D2D} and P_{MBS} are the transmission powers of VH and MBS, respectively; the channel gain $g_{Z,m} = h_{Z,m} \parallel d_{Z,m} \parallel^{-\eta}$ [19], where $Z \in \{\text{MBS, D2D}\}$, $h_{Z,m}$ is the Rayleigh fading coefficient modeled as independent and identical exponential distribution with unit mean, i.e., $h_{Z,m} \sim \exp(1)$, $\parallel . \parallel$ is the Euclidean norm, $d_{Z,m}$ is the distance between transmitter Z and RD m, and η is the path loss exponent; σ_0 is the power of additive white Gaussian noise; and σ_i is the interference transmitted by the other D2D pairs reusing the same downlink resource to transmit video i, $\forall j$, given by

$$\sigma_i = \sum_{k \in \Phi_i^{\text{VH}} \setminus \{n\}} \sum_{l \in \Phi_i^{\text{DR}} \setminus \{m\}} P_{\text{D2D}} g_{k,l}. \tag{2.7}$$

Let $R_{i,j}$ denote the rate required to offload $v_{i,j}$, for the ease of implementation, we assume that the VH located at (or nearby) the center of the circle with radius R_{D2D} is selected to offload $v_{i,j}$. Consider the worst case, when the transmission distance $d_{n,m} = R_{\text{D2D}}$, the achievable rate probability at DR m provided by VH n is given by [19]

$$\text{Pr}_{i,j}^{\text{DA}} = \Pr\{C_{n,m}^{(i)} \geq R_{i,j}\} = e^{-\xi_{i,j}\left[\lambda_{\text{MBS}}\left(\frac{P_{\text{MBS}}}{P_{\text{D2D}}}\right)^{\frac{2}{\eta}} + \lambda_i\right]}, \tag{2.8}$$

where $\xi_{i,j} = \pi (R_{\text{D2D}})^2 \Gamma\left(1 + \frac{2}{\eta}\right) \Gamma\left(1 - \frac{2}{\eta}\right) \left(2^{\frac{R_{i,j}}{W}} - 1\right)^{2/\eta}$, and λ_{MBS} is the density of the MBS.

In (2.8), λ_i is the density of the ones generating the interference. For video i, if the density of SH for sharing the downlink spectrum resources is greater than or equal to the pairing density of D2D communications, i.e., $\lambda_i^{\text{SH}} \geq \min\{\lambda_i^{\text{VH}}, \lambda_i^{\text{DR}}\}$, for offloading from the VH to the DR, no interference occurs ($\lambda_i = 0$). Otherwise, if $\lambda_i^{\text{SH}} < \min\{\lambda_i^{\text{VH}}, \lambda_i^{\text{DR}}\}$, $\lambda_i = \min\{\lambda_i^{\text{VH}}, \lambda_i^{\text{DR}}\} - \lambda_i^{\text{SH}}$. In general, λ_i is expressed as

$$\lambda_i = \max\left\{ \min\{\lambda_i^{\text{VH}}, \lambda_i^{\text{DR}}\} - \lambda_i^{\text{SH}}, 0\right\}. \tag{2.9}$$

2.3.3 Average Playback Quality and Resource Consumption

The average playback quality is considered as the objective function of the RSC optimization problem. Based on the hit and achievable rate probabilities, the average playback quality of received videos is expressed as

$$\overline{Q} = \sum_{i=1}^{I} r_i \Bigg[\underbrace{p_i \text{Pr}_i^{\text{SR}} \sum_{j=1}^{J_i} \text{Pr}_{i,j}^{\text{SA}} w_{i,j} Q_{i,j}}_{\text{SOF mode}} + \underbrace{(1 - p_i) \text{Pr}_i^{\text{DR}} \sum_{j=1}^{J_i} \text{Pr}_{i,j}^{\text{DO}} w_{i,j} Q_{i,j}}_{\text{DOF mode}} \Bigg], \tag{2.10}$$

where $\mathcal{Q}_{i,j} = 10\log_{10}\frac{255^2}{D_{i,j}}$ is the peak signal-to-noise ratio measured in dB and $\mathrm{Pr}_{i,j}^{\mathrm{DO}}$ is the successful DOF probability computed as

$$\mathrm{Pr}_{i,j}^{\mathrm{DO}} = \mathrm{Pr}_i^{\mathrm{VH}}\mathrm{Pr}_i^{\mathrm{SH}}\mathrm{Pr}_{i,j}^{\mathrm{DA}}. \tag{2.11}$$

Similar to (2.10), the average throughput and storage consumptions, constrained by the upper bounds in the RSC optimization problem, are, respectively, given by

$$\overline{R} = \sum_{i=1}^{I} r_i \left[p_i \mathrm{Pr}_i^{\mathrm{SR}} \sum_{j=1}^{J_i} \mathrm{Pr}_{i,j}^{\mathrm{SA}} w_{i,j}\mathcal{R}_{i,j} + (1-p_i)\mathrm{Pr}_i^{\mathrm{DR}} \sum_{j=1}^{J_i} \mathrm{Pr}_{i,j}^{\mathrm{DO}} w_{i,j}\mathcal{R}_{i,j} \right], \tag{2.12}$$

and

$$\overline{C} = \sum_{i=1}^{I} [p_i \mathrm{Pr}_i^{\mathrm{SR}} + (1-p_i)\mathrm{Pr}_i^{\mathrm{DR}}\mathrm{Pr}_i^{\mathrm{VH}}] \sum_{j=1}^{J_i} w_{i,j} S_{i,j}. \tag{2.13}$$

2.4 RSC Optimization Design

2.4.1 RSC Optimization Problem

Aiming to maximize the average playback quality (2.10) by finding the optimization variables p_i and $w_{i,j}$, under the constraints on throughput and storage resources, the RSC optimization problem is formulated as

$$\max_{p_i, w_{i,j}} \overline{Q}, \tag{2.14a}$$

$$\text{s.t. } 0 \le p_i \le 1, \forall i, \tag{2.14b}$$

$$\sum_{j=1}^{J_i} w_{i,j} \le 1, \forall i, \tag{2.14c}$$

$$\overline{R} \le R, \tag{2.14d}$$

$$\overline{C} \le C, \tag{2.14e}$$

where (2.14c) is to ensure that only one proper $R_{i,j}$ is selected for $v_{i,j}$, R is the limit of system throughput estimated by the statistical playback resolution of mobile devices, and C is the limit of caching storage.

2.4.2 RSC Optimization Solution with Genetic Algorithms

GA belongs to a family of adaptive heuristic algorithms based on the evolutionary principles of natural selection and genetic variation. GA is able to consider multiple peaks simultaneously without concerning whether the search space is unimodal or multimodal, but find globally (exactly or approximately) optimal results at a reasonable and flexible complexity [20–23]. Meanwhile, exhaustive algorithms can obtain exact results but are computationally expensive and time-consuming, thus are infeasible for real-world applications [3, 24]. Other promising methods, such as dynamic programming, stochastic and game learning, greedy algorithms, and heuristic strategies, can provide lower complexity, but less accuracy [23]. In this chapter, the proposed conventional GA is able to solve the RSC problem with respect to both the binary variable for spectrum sharing and the real variable for power allocation, under complicated constraints, which cannot be done by the aforementioned methods.

For the RSC optimization problem, the conventional GA is developed so that it can address the complicated constraints with respect to both real and binary optimization variables [9]. Commonly, GA supports only simple constraints like (2.14b), but not complicated ones such as (2.14c), (2.14d), and (2.14e). Therefore, we apply a penalty function [9] to address the complicated constraints by which (2.14c), (2.14d), and (2.14e) are rewritten as

$$\begin{cases} \Delta w_i = 1 - \sum_{j=1}^{J} w_{i,j} \geq 0, \forall i, \\ \Delta R = R - \overline{R} \geq 0, \\ \Delta C = C - \overline{C} \geq 0, \end{cases} \tag{2.15}$$

and the penalty function is expressed as

$$F = \lambda_1 \sum_{i=1}^{I} (\min\{0, \Delta w_i\})^2 + \lambda_2 (\min\{0, \Delta R\})^2 + \lambda_3 (\min\{0, \Delta C\})^2, \tag{2.16}$$

where λ_1, λ_2, and λ_3 are the constraint violation degrees. The individuals in the current generation are punished by violation degrees if they do not obey the constraints.

Finally, instead of directly solving (2.14), the GA can tackle the following unconstrained optimization problem

$$\max_{p_i, w_{i,j}} \overline{Q}_F = \overline{Q} - F. \tag{2.17}$$

In (2.17), when the GA gets converged, the penalty function $F = 0$ to maximize $\overline{Q}$ while satisfying all the constraints. To implement GA for solving (2.17), we define $B_{\mathrm{B}} = \sum_{i=1}^{I} J_i$ and $R_{\mathrm{B}} = I \times P$, respectively, as the number bits for all binary optimization variables $w_{i,j}$ and the number of bits for all real optimization variables p_i, here P represents the precision of a real optimization variable with 2^P quantitative levels. The general flowchart of GA is illustrated in Figure 2.4 and detailed steps are presented in Algorithm 2.1. In Algorithm 2.1, each chromosome—a binary string to represent an individual—is divided into two sub-strings. The left sub-strings, representing the binary variables ($w_{i,j}$), are directly passed to the fitness function for evaluation (i.e., computing (2.17) for ranking). In contrast, the right sub-strings must be converted to real variables (q_i) using a binary-to-real conversion operator (b2r) [25] before being passed to the fitness function evaluation for ranking. Given the number of generations N_G, GA terminates if one of the following conditions (TC) holds:

- The error of $\overline{Q}_F$ between two generations (Gen − 1 and Gen) and the penalty F are less than 10^{-3} in 10 consecutive generations.
- Gen $= N_G$.

Regarding the complexity of Algorithm 2.1, it includes memory complexity and time complexity. The memory complexity is $\mathcal{O}(N_P N_B)$, while the time complexity depends not only on N_P and N_B but also on the computations and the operators of GA such as calculating fitness values, ranking, selection, crossover, and mutation. Here, the time complexity, which is caused by mutation, can be ignored because it is executed at a very low probability (10^{-9}).

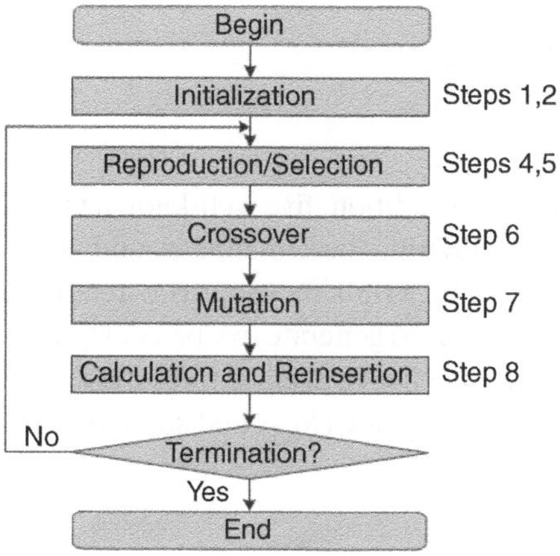

Figure 2.4 A general flowchart of GA.

Algorithm 2.1: GA for RSC Optimization Problem

Input: Parameters of system in Table 2.2
- N_{P}: Population size (6000 individuals)
- P_{G}: Generation gap (0.9)
- B_{B}: Number of bits for binary optimization variables $w_{i,j}$ (15 bits)
- R_{B}: Number of bits for real optimization variables p_i (50 bits)
- P_{C}: Crossover probability (0.6)
- P_{M}: Mutation probability (10^{-9})
- N_{G}: Number of generations (100)
- TC: Termination condition
- $Gen = 1$: Generation count

Output: $\overline{Q}_F^*(w_{i,j}^*, q_i^*)$

1: Generate N_P random strings (chromosomes), each has $N_{\mathrm{B}} = B_{\mathrm{B}} + R_{\mathrm{B}}$ bits, namely $\{X^{(z)}\}$, to represent individual z, $z = 1, 2, \ldots, N_P$
2: Compute (2.17) to have N_P fitness values $\{\overline{Q}_F(X^{(z)})\}$
3: **while** TC does not hold **do**
4:     Put $\{X^{(z)}\}$ associated with $\{\overline{Q}_F(X^{(z)})\}$ into the mating pool for ranking
5:     Select $N_{\mathrm{PG}} = N_{\mathrm{P}} \times P_{\mathrm{G}}$ individuals with higher fitness values by using stochastic universal sampling operator for breeding [25], to obtain $\{X^{(t)}\}$, $t = 1, 2, \ldots, N_{\mathrm{PG}}$
6:     Choose a pair of parents in $\{X^{(t)}\}$ to produce the offsprings by using single point crossover with probability P_{C}
7:     Mutate the offsprings with probability P_{M} to obtain $\{X'^{(t)}\}$ so that the positive genetic features probably lost in the previous steps can be recovered
8:     Compute (2.17) to have N_{PG} fitness values $\{\overline{Q}_F(X'^{(t)})\}$, then reinsert $\{X'^{(t)}\}$ and $\{\overline{Q}_F(X'^{(t)})\}$ into the current generation to get $\{X^{(z)}\}$ and $\{\overline{Q}_F(X^{(z)})\}$
9:     $Gen = Gen + 1$
10: **end while**
11: Find the highest fitness value $\overline{Q}_F^*(w_{i,j}^*, q_i^*) \in \{\overline{Q}_F(X^{(z)})\}$ in the last generation

2.5 Performance Evaluation of RSC

2.5.1 System Setting

The system parameters are listed in Table 2.2. In addition, five well-known videos including Mobile, BlowingBubbles, BasketballDrill, Basketballpass, and Racehorses are analyzed to obtain their experimental RDMs by using HM reference software version 12.0 [18]. These models enable us to generate five pairs of (γ_i, β_i). Each video is encoded into 3 versions of low rate ($j = 1$), average rate ($j = 2$), and full rate ($j = 3$). The details about (γ_i, β_i), encoding rate ($R_{i,j}$), and sizes ($S_{i,j}$) are given in Table 2.3.

2.5.2 GA Convergence Evaluation

We evaluate the convergence rate of the GA by comparing it with another heuristic algorithm, namely the binary bat algorithm (BA) [26]. We deploy the BA using

Table 2.2 Parameters setting.

Symbols	Values
$\{R_{\text{MBS}}, R_{\text{D2D}}\}$	$\{500, 15\}$m, $A = \pi R_{\text{MBS}}^2$
$\{\rho, \eta, \alpha\}$	$\{0.3, 4, 1\}$
$\{\lambda_{\text{MBS}}, \lambda_{\text{MU}}\}$	$\{1/A, 10000/A\}$
$\{\sigma_0, P_0, P_{\text{D}}\}$	$\{10^{-13}$ W, 10 W, 0.005 W$\}$
$\{W, C, R\}$	{10 MHz, 1 Gbits, 15 Mbps}
I	5 videos
J_i	3 versions, $\forall i$
$\{\lambda_1, \lambda_2, \lambda_3\}$	$\{10^2, 10^{-4}, 10^{-2}\}$

Table 2.3 Parameters of videos.

$v_{i,j}$	$\gamma_{i,j} \mid \beta_{i,j}$	$R_{i,j}$(Kbps)	$S_{i,j}$(Kbits)
$v_{1,1}$	9,806e+00 | -0.9972	1,000	11,867
$v_{1,2}$	9,806e+00 | -0.9972	2,000	23,734
$v_{1,3}$	9,806e+00 | -0.9972	3,000	35,600
$v_{2,1}$	1,644e+03 | -1.0920	50,000	33,382
$v_{2,2}$	1,644e+03 | -1.0920	100,000	66,763
$v_{2,3}$	1,644e+03 | -1.0920	170,000	113,496
$v_{3,1}$	6,566e+01 | -0.9714	10,000	1,172,340
$v_{3,2}$	6,566e+01 | -0.9714	15,000	1,758,510
$v_{3,3}$	6,566e+01 | -0.9714	20,000	2,344,680
$v_{4,1}$	7,652e+01 | -1.1530	3,000	198,680
$v_{4,2}$	7,652e+01 | -1.1530	4,000	264,906
$v_{4,3}$	7,652e+01 | -1.1530	5,300	351,000
$v_{5,1}$	1,108e+03 | -1.2060	10,000	160,410
$v_{5,2}$	1,108e+03 | -1.2060	20,000	320,820
$v_{5,3}$	1,108e+03 | -1.2060	28,300	453,960

the same penalty function and population size (N_P) as the GA. To enable the BA to handle real optimization variables for caching probabilities, we set $P = 10$ bits to represent each real variable p_i, mapped to the probability range [0, 1]. As shown in Figure 2.5, the GA converges after 30 generations, when the penalty function (F in [2.16] and [2.17]) reaches zero, ensuring that all constraints are satisfied. At the same time, the mean of fitness values computed over the selected individuals in step 5 of Algorithm 2.1 approaches the highest fitness value, i.e., the best

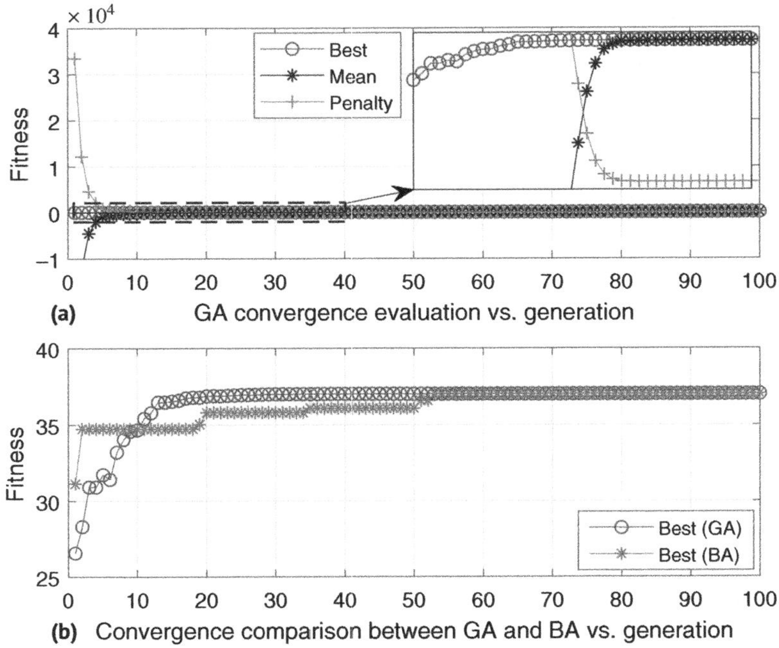

Figure 2.5 Convergence rate evaluation. *Source*: Reproduced from [1] / IEEE.

individual. In the last generation, all the individuals possess the same best genetic features in accordance with the maximum value of average playback quality. In contrast, the BA requires more than 50 generations to converge. These results demonstrate the effectiveness of the GA in solving the RSC optimization problem.

Furthermore, the population size (N_P) significantly impacts the performance of the GA: a larger N_P yields more accurate optimal results but requires greater computational time and memory. To select a proper number of N_P, we consider different values of N_P ranging from 1000, 3000,..., to 10000, and then repeat the GA 100 tries for each N_P to obtain 100 approximate or exact optimal results. For each N_P, we compute the minimum (Min), average (Ave), and maximum (Max) values of 100 results as shown in Figure 2.6. The results in Figure 2.6 show that the proper value of N_P is 6000, happening at the elbow point of the Ave. From the elbow point, even though we increase N_P, the Ave is not significantly improved, but it incurs substantially higher time and memory requirements. It is noted in Figure 2.6 that the Max is equivalent to the exact optimal results held if the N_P is large enough, i.e., $N_P = 2000$.

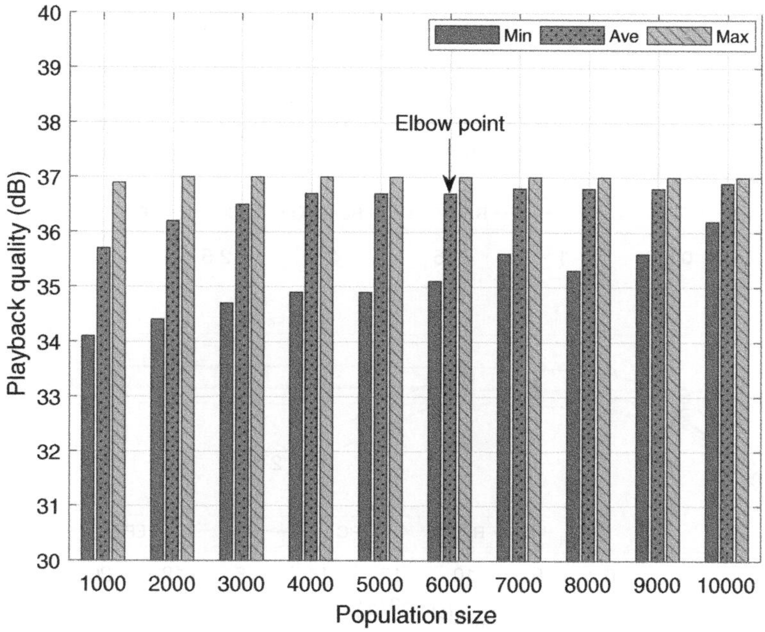

Figure 2.6 Playback quality versus N_P.

2.5.3 RSC Evaluation

To evaluate the performance of RSC, it is compared with three other schemes, including full-rate caching (FRC), low-rate caching (LRC), and equal-probability caching (EPC). In FRC, among three versions of each video, the ones encoded at full rates for the highest reconstructed qualities are likely to be cached, and R is relaxed to 30 Mbps. Meanwhile, in LRC, the ones encoded at low rates for the lowest reconstructed qualities are likely to be cached, and the constraints are kept the same with RSC. In EPC, the optimal caching probabilities are found equally for all videos, and R is relaxed to 20 Mbps. In fact, the FRC and LRC schemes represent the conventional methods that do not consider selecting the optimal encoding rate of each video for caching, and the EPC represents an important benchmark for insightful evaluation. The detailed performance evaluation is presented in the sequel.

In Figure 2.7a,b, the performance of all the schemes is evaluated versus α and λ_{MU}, respectively. With respect to α, the RSC generally outperforms the FRC, except when α is sufficiently large. As α increases, both RSC and FRC tend to cache

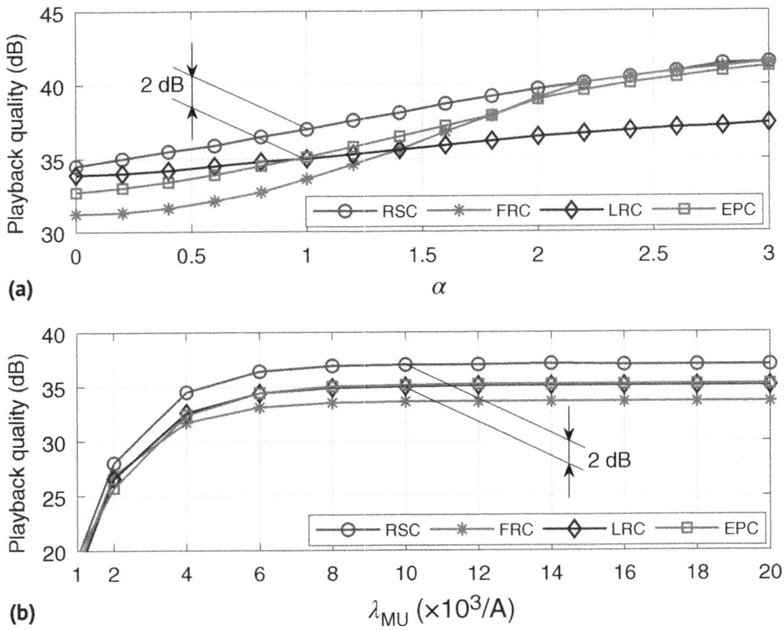

Figure 2.7 Playback quality versus (a) α and (b) λ_{MU}. *Source*: Reproduced from [22] / IEEE.

the most popular videos with higher probabilities and full rates, explaining this behavior. In doing so, the FRC consumes significantly more bandwidth than the RSC, i.e., see Figure 2.9b. Moreover, the RSC consistently provides higher playback quality than the LRC and EPC. With respect to λ_{MU}, the results in Figure 2.7b indicate that as MU density increases, the RSC outperforms the other schemes. It is certain that the playback quality of all schemes saturates, as the encoding rate, storage, and throughput are physically constrained.

Figure 2.8 illustrates the performance of all the schemes versus R_{D2D}. In Figure 2.8a, when R_{D2D} is too short, the performance rapidly decreases since there are no MUs for caching and offloading. If R_{D2D} increases, more MUs for caching and offloading within a reasonable transmission radius make the performance significantly improved. However, when R_{D2D} is too long, the DOF mode cannot be established and is therefore replaced by the SOF mode. As a result, the performance saturates, being served only by the SOF mode, after approaching a certain maximum value. The maximum values are obtained when R_{D2D} meets the number of MUs for caching and offloading. In comparison, the RSC generally

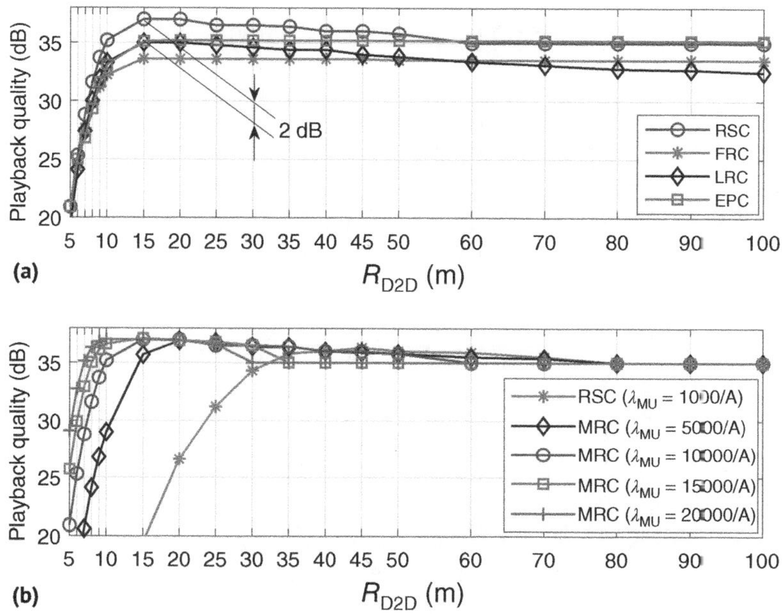

Figure 2.8 Playback quality of RSC versus R_{D2D}: (a) performance comparison among different schemes; (b) playback quality of RSC under different MU densities λ_{MU}. *Source*: Reproduced from [1] / IEEE.

outperforms the EPC and consistently outperforms the FRC and LRC. Note that the RSC cannot surpass the EPC under saturation, as the EPC benefits from relaxed throughput constraints (R), as shown in Figure 2.9b. In addition, in Figure 2.8b, the performance of RSC is further investigated by changing the density of MUs (λ_{MU}). Particularly, the maximum values of RSC move from left (high λ_{MU}) to right (low λ_{MU}). This enables us to find the optimal value of cluster radius (R_{D2D}) associated with a given density of MUs for the highest playback quality.

In Figure 2.9, the consumption of storage and throughput resources in all schemes is evaluated. We can see that the proposed RSC is able to conserve the resources compared to the FRC, LRC, and EPC. Although the RSC consumes slightly more storage than the EPC and slightly more bandwidth than the LRC, it significantly improves playback quality, especially in scenarios with dense MUs. Importantly, the resource constraints in the RSC optimization problem can be adjusted to serve the RUs flexibly under various video offloading scenarios in dense D2D networks.

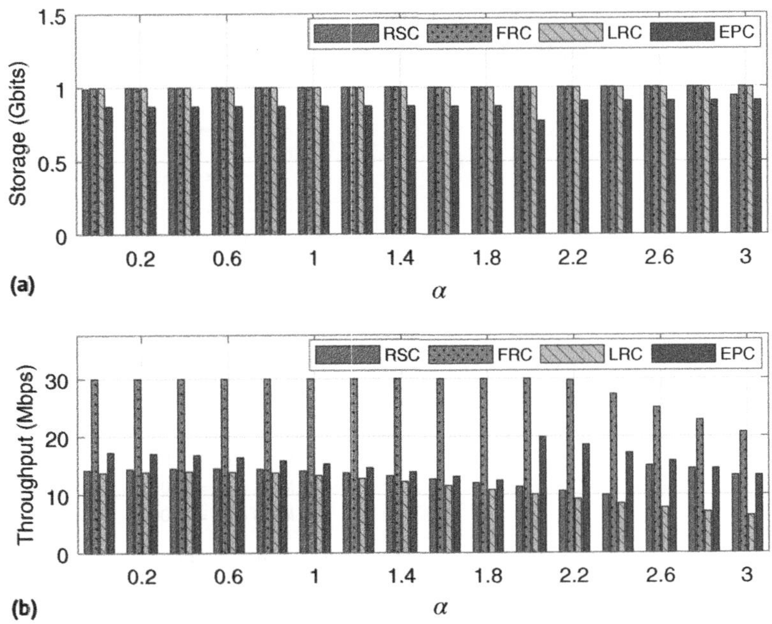

Figure 2.9 (a) Storage and (b) throughput consumption versus α. *Source*: Reproduced from [1] / IEEE.

2.6 Conclusion

In this chapter, we have proposed the RSC solution for video offloading in dense D2D networks. The RSC not only exploited the probabilistic caching and underlay D2D communication with spectrum sharing techniques, but also utilized the popularity and RDM for videos and homogeneous Poisson Point Process model for MU distribution. To maximize the playback quality, the RSC optimization problem was formulated to find the optimal caching probability and encoding rate selection of each video under the constraints on the system throughput and caching storage resources. Simulation results demonstrated the benefits of the proposed RSC solution compared to other schemes in terms of playback quality and resource utilization and provided useful insights into the caching optimization system design for local VAS in dense D2D networks.

References

1 Q.-N. Tran, N.-S. Vo, T.-M. Phan, T. C. Lam, and A. Masaracchia. Multi-rate probabilistic caching optimised video offloading in dense D2D networks. *IEEE Commun. Lett.*, 27(4):1240–1244. 2023.

2 D. V. Huynh, S. R. Khosravirad, A. Masaracchia, O. A. Dobre, and T. Q. Duong. Edge intelligence-based ultra-reliable and low-latency communications for digital twin-enabled metaverse. *IEEE Wirel. Commun. Lett.*, 11(8):1733–1737. 2022

3 N.-S. Vo, M.-P. Bui, P.Q. Truong, C. Yin, and A. Masaracchia. Multi-tier caching and resource sharing for video streaming in 5G ultra-dense networks. *IEEE Commun. Lett.*, 24(7):1500–1504. 2020.

4 J. Ma, L. Liu, B. Shang, S. Jere, and P. Fan. Performance analysis and optimization for layer-based scalable video caching in 6G networks. *IEEE/ACM Trans. Netw.*, 31(4):1494–1506. 2023.

5 T. C. Lam, N.-S. Vo, T.-H. Nguyen, T.-M. Phan, and D.-T. Huynh. Genetic algorithms for storage- and energy-aware caching and trajectory optimisation problem in UAV-assisted content delivery networks. In *Proc. EAI Inter. Industrial Netw. and Intell. Syst. Conf.*, pages 34–50, Ho Chi Minh City, Vietnam, Aug. 2023.

6 N.-S. Vo, T.-M. Phan, M.-P. Bui, X.-K. Dang, V. T. Nguyen, and C. Yin. Social-aware spectrum sharing and caching helper selection strategy optimized multicast video streaming in dense D2D 5G networks. *IEEE Syst. J.*, 15(3):3480–3491. 2021.

7 Q.-N. Tran, N.-S. Vo, T.-M. Phan, M.-P. Bui, M.-N. Nguyen, and A. Kortun. Downlink resource allocation maximized video delivery capacity over multi-hop multi-path in dense D2D 5G networks. In *Proc. Inter. Conf. Recent Advances on Signal Processing, Telecommun. & Computing*, pages 72–76, Ha Noi, Vietnam, Aug. 2020.

8 Q.-N. Tran, N.-S. Vo, Q.-A. Nguyen, M.-P. Bui, T.-M. Phan, V.-V. Lam, and A. Masaracchia. D2D multi-hop multi-path communications in B5G networks: A survey on models, techniques, and applications. *EAI Endorsed Trans. Industrial Netw. and Intell. Syst.*, 7(25):1–12. 2021.

9 Q.-N. Tran, N. S. Vo, M.-P. Bui, T.-M. Phan, Q.-A. Nguyen, and T. Q. Duong. Spectrum sharing and power allocation optimised multihop multipath D2D video delivery in beyond 5G networks. *IEEE Trans. Cogn. Commun. and Netw.*, 8(2):919–930. 2022.

10 Z. Chen, N. Pappas, and M. Kountouris. Probabilistic caching in wireless D2D networks: Cache hit optimal versus throughput optimal. *IEEE Commun. Lett.*, 21(3):584–587. 2017.

11 J. Ma, L. Liu, B. Shang, and P. Fan. Cache-aided cooperative device-to-device (D2D) networks: A stochastic geometry view. *IEEE Trans. Commun.*, 67(11): 7444–7455. 2019.

12 R. Amer, H. Elsawy, M. M. Butt, E. A. Jorswieck, M. Bennis, and N. Marchetti. Optimized caching and spectrum partitioning for D2D enabled cellular systems with clustered devices. *IEEE Trans. Commun.*, 68(7):4358–4374. 2020.

13 R. Amer, M. Baza, T. Salman, M. M. Butt, A. Alhindi, and N. Marchetti. Optimizing joint probabilistic caching and channel access for clustered D2D networks. *J. Commun. and Netw.*, 23(6):433–441. 2021

14 Y. Meng, Z. Zhang, and Y. Huang. Cache- and energy harvesting-enabled D2D cellular network: Modeling, analysis and optimization. *IEEE Trans. Green Commun. and Netw.*, 5(2):703–713. 2021.

15 M. Naslcheraghi, C. Wette, J.-F. Frigon, and B. Sansó. Probabilistic analysis of operating modes in cache-enabled full-duplex D2D networks. *IEEE Trans. Veh. Technol.*, 71(6):6624–6635. 2022.

16 X. Song, Y. Geng, X. Meng, J. Liu, W. Lei, and Y. Wen. Cache-enabled device to device networks with contention-based multimedia delivery. *IEEE Access*, 5:3228–3239. 2017.

17 J. Ma, L. Liu, H. Song, R. Shafin, B. Shang, and P. Fan. Scalable video transmission in cache-aided device-to-device networks. *IEEE Trans. Wirel. Commun.*, 19(6):4247–4261. 2020.

18 HEVC. *HM Reference Software Version 12.0*. http://hevc.hhi.fraunhofer.de. 2013.

19 A. Bhardwaj and S. Agnihotri. Energy- and spectral-efficiency trade-off for D2D-multicasts in underlay cellular networks. *IEEE Wirel. Commun. Lett.*, 7(4):546–549. 2018.

20 W. Xiang, C. Zhu, C. K. Siew, Y. Xu, and M. Liu. Forward error correction-based 2-D layered multiple description coding for error-resilient H.264 SVC video transmission. *IEEE Trans. Circuits Syst. Video Technol.*, 19(12):1730–1738. 2009.

21 X. Du, N.-S. Vo, T. Q. Duong, and L. Shu. Joint replication density and rate allocation optimization for VoD systems over wireless mesh networks. *IEEE Trans. Circuits Syst. Video Technol.*, 23(7):1260–1273. 2013.

22 N.-S. Vo, T. Q. Duong, T. D. Hoang, and A. Kortun. Optimal video streaming in dense 5G networks with D2D communications. *IEEE Access*, 6:209–223. 2017.

23 M.-P. Bui, N.-S. Vo, T.-V. Truong, T.-H. Nguyen, N. V. Nguyen, and C. Yin. Genetic algorithms for multi-tier caching and resource sharing optimized video streaming in 5G ultra-dense networks. In *Proc. Inter. Conf. Recent Advances on Signal Processing, Telecommun. & Computing*, pages 66–71, Ha Noi, Vietnam, Aug. 2020.

24 M.-P. Bui, N.-S. Vo, S. Q. Nguyen, and Q.-N. Tran. Social-aware caching and resource sharing maximized video delivery capacity in 5G ultra-dense networks. *Springer Mob. Netw. Appl.*, 25:2037–2049. 2019.

25 A. Chipperfield, P. Fleming, H. Pohlheim, and C. Fonseca. *Genetic Algorithm TOOLBOX for Using with Matlab Ver 1.2 Users Guide*. University of Sheffield, 1994.

26 S. Mirjalili, S. M. Mirjalili, and X.-S. Yang. Binary bat algorithm. *Neural Comput. Appl.*, 25:663–681. 2014.

3

Spectrum Sharing and Power Allocation for Video Transmission in Multi-hop Multi-path D2D Networks

3.1 Introduction

In dense 6G networks, an ever-increasing number of MUs requesting VAS generates enormous data volumes at extremely high rates, causing severe congestion in the backhaul links of MBSs. Meanwhile, investing in high-capacity backhaul links is costly, yet numerous resources in mobile devices, such as storage, spectrum, and energy, remain underutilized. This indicates an urgent need to design a proper offloading network that (i) exploits the mobile devices' resources while remaining low cost for achieving a better quality of service (QoS) of VAS and (ii) mitigates the congestion situations at the backhaul links of MBS.[1]

One of the most efficient offloading networks is D2D networks. They utilize all the available storage, spectrum, and energy resources of mobile devices for directly offloading (unicasting or multicasting) the videos to the requesting MUs (RUs) over D2D communications [2–14]. Especially, deep reinforcement learning and online learning algorithms have been studied to assist D2D networks in capturing the varying environments to satisfy the RUs [15, 16]. However, direct D2D offloading is limited in that within a dense network, it cannot leverage the videos cached in the MUs—referred to as video helping MUs (VHs) and located more than one D2D hop from the RU—nor can it distribute energy consumption across the network via MHMP communications.

The advantages of MUs in a dense network have been studied for leveraging spectrum and energy resources in MHMP D2D communications, which can extend coverage from the MBS [17–22]. Caching and the content's features, such as popularity, size, and VAS/non-VAS, have not been considered. Some studies have focused on MHMP D2D communications by utilizing the cached videos in the VHs and the popularity of videos, which are effectively applied to VAS [23–31].

1 This chapter has been published partly in [1].

However, most studies have not integrated the joint spectrum and power allocation into MHMP D2D and conventional cellular communications, preventing the VHs and MBS from cooperatively serving the RUs with the highest video delivery capacity and low transmission power consumption.

In this chapter, we propose a joint spectrum sharing and power allocation (SPA) solution for VAS over underlay MHMP D2D communications in dense 6G networks. The SPA exploits the available videos, which have been cached in the VHs. It also takes into account the popularity of videos to serve the RU more efficiently. Energy resources of the transmitters (Tx) and the receivers (Rx)[2] of D2D hops are utilized to forward the requested videos from the VH to the RU. Spectrum resources of the spectrum helping MUs (SHs) are reused to establish the D2D communications between the Tx and the Rx in each path.

An SPA optimization problem is formulated for finding the optimal values of (i) spectrum sharing pairs between the SHs and D2D hops, (ii) powers allocated to the Txs, and (iii) transmission modes between MHMP D2D communications from the VHs and conventional cellular communications from the MBS. The total power of the Tx and the target signal-to-interference-plus-noise ratio (SINR) of the SH are considered as the constraints to conserve the power consumption and limit the interference impact on the SH when sharing the spectrum resource with the Tx. Genetic algorithms (GAs) are applied to solve the SPA optimization problem feasibly. As a result, the MBS can optimally decide whether to deliver the requested videos itself via normal cellular communications or to assign the VH to do so via MHMP D2D communications, to maximize video delivery capacity. Simulation results are provided to demonstrate the GA's feasibility and the benefits of the SPA method against other conventional ones.

The rest of this chapter is organized as follows. In Section 3.2, we introduce the system model for VAS over underlay MHMP D2D communications in collaboration with the MBS. The SPA model is formulated in Section 3.3, followed by the SPA optimization problem and its solution using GA in Section 3.4. Section 3.5 evaluates the performance of the GA and SPA method through simulation. Finally, we conclude the chapter in Section 3.6.

3.2 SPA Model for Video Transmission in MHMP D2D Networks

We consider a system model for VAS over underlay MHMP D2D communications as shown in Figure 3.1. The system, which is designed to assist the MBS, has I VHs, N SHs, and one RU. VH i has already cached video i, $i = 1, 2, \ldots, I$. The MBS can

2 In MHMP D2D communications, an Rx of the current D2D hop is the Tx of the next D2D hop.

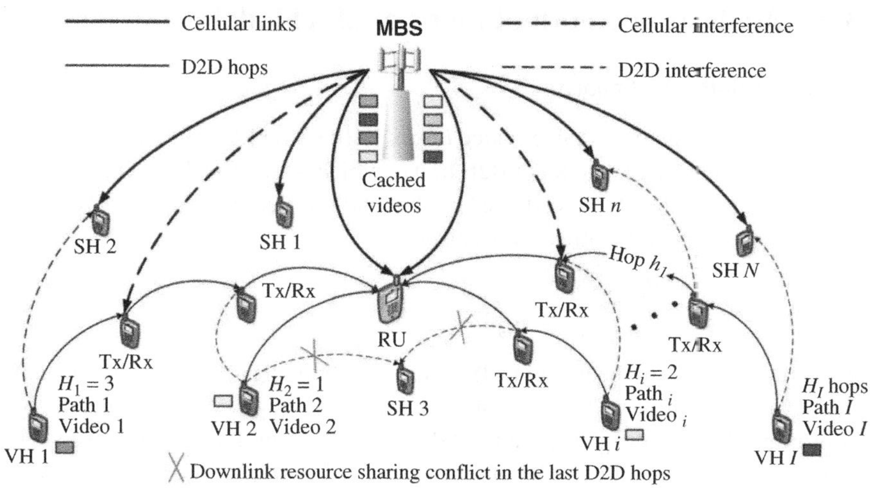

Figure 3.1 VAS over MHMP D2D communications.

cache I videos. There are I paths from I VHs to the RU. Path i has H_i D2D hops, each of a pair Tx–Rx h_i, $h_i = 1, 2, ..., H_i$. The SUs share their downlink spectrum resources with the Txs to establish the MHMP D2D communications by applying the interference-aware multi-hop path selection scheme [20]. To fully utilize the downlink spectrum resources, an SH is allowed to share with up to I Txs in I paths, but with only one Tx in each path. To avoid downlink resource sharing conflict, there must be I SHs to share their downlink resources separately with I Txs in the last D2D hops, which are directly connected to the RU.

When acknowledging that the RU requests for a number of videos, the MBS operates the VAS over MHMP D2D communications by collecting the system information, such as characteristics of wireless channels related to the VHs, SHs, and D2D hops; popularity pattern of cached videos; and power capability of the Txs. The SPA optimization problem is formulated and solved for the optimization variables of (i) downlink resource sharing index ($d_{n,h_i} \in \{0, 1\}$ to indicate whether SH n, $n = 1, 2, ..., N$, agrees to share its downlink resource with Tx h_i in the path i ($d_{n,h_i} = 1$) or not ($d_{n,h_i} = 0$), (ii) transmission power P_{h_i} allocated to Tx h_i, and (iii) transmission mode index ($b_i \in \{0, 1\}$) to make a decision that whether path i is selected ($b_i = 1$) or not ($b_i = 0$, i.e., MBS is selected) for sending video i to the RU. Accordingly, the video capacity delivered to the RU is maximized under the constraints on the total power of the Txs and the target SINR of the SHs.

3.3 Analysis and Formulations of SPA Model

3.3.1 MHMP D2D Capacity

To derive the capacity to deliver video i from VH i to the RU over path i, we need to compute the SINR at the Rx of D2D hop h_i which reuses the downlink resource shared by SH n. The SINR is affected by the total interference caused by the MBS and the other Txs in the other paths, given by

$$\gamma_{n,h_i} = \frac{d_{n,h_i} P_{h_i} g_{h_i}}{\sigma_0 + P_{\mathrm{SH}} g_{\mathrm{MBS},h_i} + \sigma_{h_i}}, \tag{3.1}$$

where P_{h_i} and g_{h_i} are the power of Tx h_i and the channel gain of D2D hop h_i, σ_0 is the power of additive white Gaussian noise, P_{SH} and g_{MBS,h_i} are the transmission power and the channel gain from the MBS to the SHs and Rx h_i, respectively. The channel gains follow the model of an exponential power fading coefficient with unit mean, i.e., $\sim \exp(1)$, multiplied by a standard power-law path loss function with pathloss exponent η [32]. In addition, the interference I_{h_i} caused by the Txs in the other paths to Rx h_i in path i is expressed as

$$\sigma_{h_i} = \sum_{l=1,l\neq i}^{I} \sum_{h_l=1}^{H_l} d_{n,h_l} P_{h_l} g_{h_l,h_i}, \tag{3.2}$$

where P_{h_l} is the Tx power from D2D hop l in path l $(l \neq i)$ and g_{h_l,h_i} is the channel gain between Tx h_l in the path l and Rx h_i in path i.

In path i, the best D2D hop h_i with the highest SINR γ_{h_i} is established by the optimal values of P_{h_i} and d_{n,h_i}, and thus we have

$$\gamma_{h_i} = \max\{\gamma_{n,h_i}, n = 1, 2, \ldots, N\}. \tag{3.3}$$

The SINR γ_i of path i represented by the worst D2D hop with the lowest SINR is expressed as

$$\gamma_i = \min\left\{\gamma_{h_i}, h_i = 1, 2, \ldots, H_i\right\}. \tag{3.4}$$

The capacity to deliver video i from VH i to the RU over path i is given by

$$C_i = W \log_2(1 + \gamma_i), \tag{3.5}$$

where W is the system bandwidth.

3.3.2 Normal Cellular Capacity

When the path from a VH to the RU cannot be established due to the lack or the limited target SINR of the SHs, or when the capacity provided by the VH is lower than that provided by the MBS, the RU can alternately receive the requested video from the MBS over normal cellular communications. The capacity at the RU delivered from the MBS is expressed as

$$C_{\mathrm{MBS}} = W \log_2 \left(1 + \frac{P_{\mathrm{RU}} g_{\mathrm{RU}}}{\sigma_0}\right), \tag{3.6}$$

where P_{RU} and g_{RU} are, respectively, the transmission power and the channel gain from the MBS to the RU.

3.3.3 Average Delivery Capacity at RU

So far, the average capacity of all requested videos delivered from the VHs or the MBS to the RU is given by

$$\overline{C} = \sum_{i=1}^{I} r_i [b_i C_i + (1 - b_i) C_{\mathrm{MBS}}], \tag{3.7}$$

where r_i is the popularity of video i (2.1) following Zipf-like distribution [33].

3.3.4 SINR at SHs

In underlay MHMP D2D communications, the Txs of D2D hops, which reuse the same downlink resources with the SHs, generate interference to the SHs. To ensure high QoS for the SHs in the SPA optimization problem, the SINR at SH n must be considered as [32]

$$\gamma_n = \frac{P_{\mathrm{SH}} g_{\mathrm{MBS},n}}{\sigma_0 + \sum_{i=1}^{I} \sum_{h_i=1}^{H_i} d_{n,h_i} P_{h_i} g_{n,h_i}}, \tag{3.8}$$

where $g_{\mathrm{MBS},n}$ and g_{n,h_i} are the channel gains from MBS and Tx h_i to SH n.

3.4 SPA Optimization Design

3.4.1 SPA Optimization Problem

Considering the objective function (3.7), together with the constraints on d_{n,h_k} for fully exploiting the downlink resources shared by the SHs, the target SINR (γ_0) for guaranteeing the QoS of the SHs, and P for limiting the total transmission power of the Txs, the SPA optimization problem is formulated as follows:

$$\max_{d_{n,h_i}, P_{h_i}, b_i} \overline{C}, \tag{3.9a}$$

$$\text{s.t.} \sum_{h_i=1}^{H_i} d_{n,h_i} \leq 1, n = 1, 2, \ldots, N, i = 1, 2, \ldots, I, \tag{3.9b}$$

$$\sum_{n=1}^{N} d_{n,h_i} \leq 1, h_i = 1, 2, \ldots, H_i, i = 1, 2, \ldots, I, \tag{3.9c}$$

$$\sum_{i=1}^{I} d_{n,H_i} \leq 1, n = 1, 2, \ldots, N, \tag{3.9d}$$

$$\sum_{i=1}^{I} \sum_{h_i=1}^{H_i} d_{n,h_i} P_{h_i} g_{n,h_i} \leq \frac{P_{\mathrm{SH}} g_{\mathrm{MBS},n}}{\gamma_0} - \sigma_0, \quad n = 1, 2, \ldots, N, \tag{3.9e}$$

$$P_{h_i} \geq 0, i = 1, 2, \ldots, I, h_i = 1, 2, \ldots, H_i, \tag{3.9f}$$

$$\sum_{i=1}^{I} \sum_{h_i=1}^{H_i} P_{h_i} \leq P, \tag{3.9g}$$

where the constraints (3.9b), (3.9c), and (3.9d) are to ensure the downlink resource sharing rules of the SHs in I paths, each path, and in the last D2D hops directly connected to the RU, respectively. The constraint (3.9e) is to guarantee the QoS of the SHs, derived by letting $\gamma_n \geq \gamma_0$. In addition, the constraints (3.9f) and (3.9g) are used to limit the total transmission power consumption of the Txs.

3.4.2 SPA Optimization with GA

Similar to Chapter 2, we apply GA [34] to solve the above constrained SPA optimization problem (3.9). To do so, we use a penalty function [35] to convert (3.9) to an unconstrained one by rewriting the complicated constraints (3.9b), (3.9c), (3.9d), (3.9e), and (3.9g) as below

$$\begin{cases} \Delta d_{n,i} = 1 - \sum_{h_i=1}^{H_i} d_{n,h_i} \geq 0, n = 1, 2, \ldots, N, i = 1, 2, \ldots, I, \\ \Delta d_{i,h_i} = 1 - \sum_{n=1}^{N} d_{n,h_i} \geq 0, i = 1, 2, \ldots, I, h_i = 1, 2, \ldots, H_i, \\ \Delta d_n = 1 - \sum_{i=1}^{I} d_{n,H_i} \geq 0, n = 1, 2, \ldots, N, \\ \Delta \gamma_n = \frac{P_{\mathrm{SH}} g_{\mathrm{MBS},n}}{\gamma_0} - \sigma_0 - \sum_{i=1}^{I} \sum_{h_i=1}^{H_i} d_{n,h_i} P_{h_i} g_{n,h_i} \geq 0, \\ n = 1, 2, \ldots, N, \\ \Delta P = P - \sum_{i=1}^{I} \sum_{h_i=1}^{H_i} P_{h_i} \geq 0. \end{cases} \tag{3.10}$$

We then derive the penalty function as

$$F = \lambda_1 \sum_{n=1}^{N} \sum_{i=1}^{I} \left(\min\{0, \Delta d_{n,i}\}\right)^2 + \lambda_2 \sum_{i=1}^{I} \sum_{h_i=1}^{H_i} \left(\min\{0, \Delta d_{i,h_i}\}\right)^2$$
$$+ \lambda_3 \sum_{n=1}^{N} \left(\min\{0, \Delta d_n\}\right)^2 + \lambda_4 \sum_{n=1}^{N} \left(\min\{0, \Delta \gamma_n\}\right)^2 + \lambda_5 \left(\min\{0, \Delta P\}\right)^2, \tag{3.11}$$

where $\lambda_1, \lambda_2, \lambda_3, \lambda_4$, and λ_5 are the constraint violation degrees. The individuals in the current generation are punished by violation degrees if they do not obey the constraints.

Finally, the GA is capable of solving the following unconstrained SPA optimization problem:

$$\max_{d_{n,h_i}, P_{h_i}, b_i} \quad \overline{C}_F = \overline{C} - F. \tag{3.12}$$

The detailed GA for solving the unconstrained SPA problem is presented in Algorithm 3.1 whose termination condition TC is the same as that in Algorithm 2.1.

3.5 Performance Evaluation of SPA

3.5.1 Parameter Setting

The system parameters are listed in Table 3.1. P is the total transmission power of the Txs consumed in the downlink spectrum sharing without power allocation (SSA) scheme [36]. In SSA, the transmission power of each Tx is generated by following a random uniform distribution in the range of $[0.001, 0.01]$ W.

3.5.2 GA Convergence Evaluation

To evaluate the feasibility of GA, we validate the convergence of GA within $N_G = 100$ generations as shown in Figure 3.2. The result demonstrates that GA quickly starts to converge after 50 generations when "Penalty" decreases to zero (i.e., $F = 0$) to ensure that all the constraints are satisfied. At the same time, the "Mean" approaches and eventually equals the "Best" in the final generations, indicating that all individuals have converged to the same optimal genetic features to maximize the objective function (3.7).

Algorithm 3.1: GA for PSA Optimization Problem

Input: System parameters in Table 3.1

N_P: Population size (10,000 individuals)

P_G: Generation gap (0.9)

$PRECI_B$: 1 bit for each binary variable of d_{n,h_i} and b_i

$PRECI_R$: 10 bits for the precision of each real variable of P_{h_i}

$B = N \times \sum_{i=1}^{I} H_i + I$: Number of binary variables of d_{n,h_i} and b_i

$R = \sum_{i=1}^{I} H_i$: Number of real variables of P_{h_i}

$B_B = PRECI_B \times B$: Length of the left sub-string

$R_B = PRECI_R \times R$: Length of the right sub-string

$N_B = B_B + R_B$: Length a chromosome or a binary string representing an individual in the potential solution set $\{X^{(z)}\}$, $z = 1, 2, \ldots, N_P$

P_C: Crossover probability (0.6)

P_M: Mutation probability (10^{-9})

N_G: Number of generations (100)

TC: Termination condition

$Gen = 1$: Generation count

Output: $\overline{C}_F^*(d_{n,h_i}^*, b_i^*, P_{h_i}^*)$

1: Generate N_P random strings, each of length N_B bits, namely $\{X^{(z)}\}$, to represent individual z

2: Divide each string $\{X^{(z)}\}$ into two substrings $\{[X_1^{(z)} X_2^{(z)}]\}$, $X_1^{(z)}$ has B_B bits and $X_2^{(z)}$ has R_B bits

3: Compute N_P fitness values $\overline{C}_F(X_1^{(z)}, \text{b2r}(X_2^{(z)}))$ with respect to individual z, b2r(.) is a binary string to real variable conversion operator [34]

4: **while** TC does not hold **do**

5: Put $\{[X_1^{(z)} X_2^{(z)}]\}$ associated with $\overline{C}_F(X_1^{(z)}, \text{b2r}(X_2^{(z)}))$ into the mating pool for ranking

6: Select $N_{PG} = N_P \times P_G$ better individuals with higher fitness values for breeding by using stochastic universal sampling operator [34]

7: Select a pair of parents to generate a pair of offsprings by using double point crossover operator done with a crossover probability P_C [34]

8: Mutate the offsprings with a mutation probability P_M to recover good genetic materials

9: Evaluate the fitness values of the offsprings and reinsert them into the present generation

10: $Gen = Gen + 1$

11: **end while**

12: Find the best fitness value $\overline{C}_F^*(d_{n,h_i}^*, b_i^*, P_{h_i}^*) \in \overline{C}_F(X_1^{(z)}, \text{b2r}(X_2^{(z)}))$ in the last generation

---

3.5.3 SPA Evaluation

3.5.3.1 Capacity Performance

To evaluate the capacity performance of the SPA solution, we compare it to other benchmarks, including SSA, average (AVE), minimum (MIN), and only MBS (MBS) schemes. In SSA [36], only downlink spectrum sharing allocation is optimized, while the transmission powers of the Txs are fixed. In AVE and MIN, we generate a large set of individuals that satisfy the constraints, and then compute the average value of $\overline{C}$ per individual and find the minimum value of $\overline{C}$ in the set. And in MBS, the RU is served by only the MBS over conventional cellular communications.

Table 3.1 Parameters setting.

Symbols	Values
N	5 SHs
I	3 VHs/paths/videos
$\{H_i\}$	$\{1, 2, 3\}$
W	10 MHz
$P_{\mathrm{SH}}, P_{\mathrm{RU}}$	10 W
σ_0	10^{-13} W
η	4 (path loss exponent)
γ_0	5 dB
α	1 (2.1)
$\{\lambda_1, \lambda_2, \lambda_3, \lambda_4, \lambda_5\}$	$\{10, 10^6, 1, 10^{26}, 10^8\}$
Distances	**Specifications**
MBS to RU	500 m
MBS to SHs/Rxs	Randomly uniform distributed in [500, 1000] m
D2D hops	Randomly uniform distributed in [25, 35] m
D2D interference links	Randomly uniform distributed in [50, 150] m

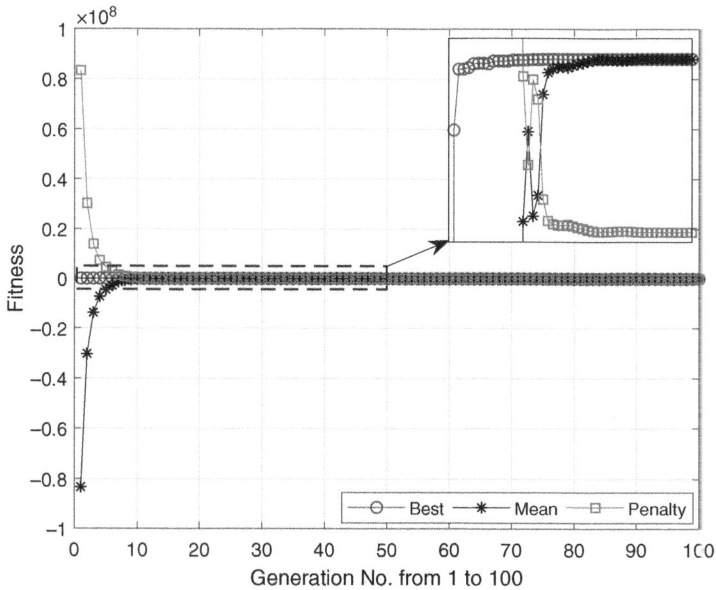

Figure 3.2 Convergence rate of GA. *Source*: Reproduced from [1] / IEEE.

The capacity performance of all schemes is first evaluated versus the exponent coefficient of skewed popularity among different videos α. As shown in Figure 3.3, when α increases, fewer videos are much more popular than the rear ones. The optimal results of PSA and SSA focus on the few most popular videos to serve the RU better. Obviously, the AVE follows the same increasing pattern as the PSA and SSA, whereas the MIN and MBS do not. The reason is that both MIN and MBS have no optimal results with respect to α, and thus the capacity does not change. Thanks to the joint optimization solution of downlink resource sharing and transmission power allocation, the SPA outperforms the other SSO, AVE, MIN, and MBS.

Next, the capacity performance of all schemes is evaluated versus the target SINR γ_0 as shown in Figure 3.4. Increasing γ_0 reduces the interference from the Txs of D2D hops, thereby ensuring higher QoS for the SHs. It means that the SHs do not want to share their downlink resources anymore. As a result, the video delivery capacity decreases because of fewer MHMP D2D communications established. When γ_0 is high enough, i.e., $\gamma_0 = 25$ db, the performance of SPA, SSA, and AVE becomes that of MIN and MBS since there are no MHMP D2D communications. In system design, we can select a proper value of γ_0 such that SPA can simultaneously guarantee the SHs high QoS and provide the RU with the highest video delivery capacity. It can be seen in Figure 3.4 that $\gamma_0 = 5$ db is proper to make the SPA significantly surpass the others.

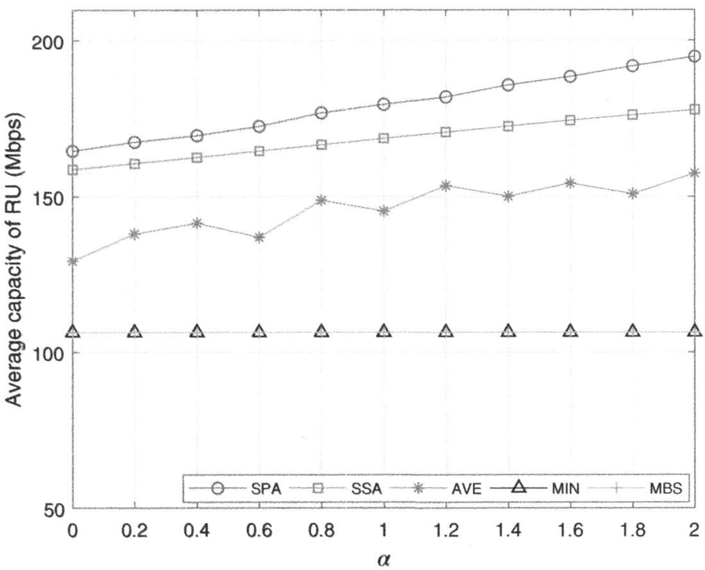

Figure 3.3 Video capacity delivered to RU versus α. *Source*: Reproduced from [1] / IEEE.

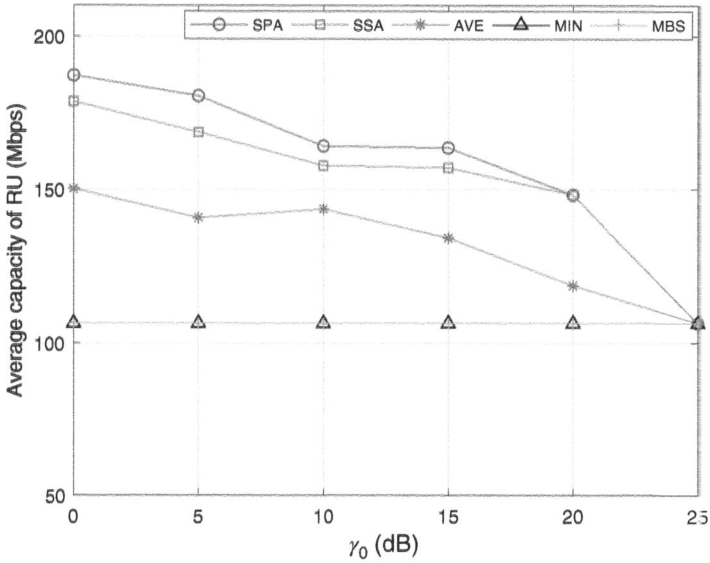

Figure 3.4 Video capacity delivered to RU versus γ_0. *Source*: Reproduced from [1] / IEEE.

Then, we evaluate the capacity performance of all schemes versus the number of SHs N as shown in Figure 3.5. It is clear that the more SHs the system has, the more available downlink spectrum resources the Txs reuse to establish better MHMP D2D communications. Thus, if N is too small (e.g., $N = 1$), the performance of all schemes is identical to that of MIN and MBS, because with only one SH, MHMP D2D communications from the VHs to the RU cannot be established, offering no improvement over conventional cellular communications from the MBS to the RU. Increasing N enables the SPA and SSA to have more opportunities for sharing the downlink spectrum resources, and thus achieving higher video delivery capacity. However, it is noted that the video delivery capacity of the SPA and SSA is likely to get saturated if we continue to increase the number of SHs (e.g., $N \geq 3$). The reason is that the more SHs we add to the system, the less possibility the new SHs added are better than the existing ones. So, for efficient system design, we carefully select the number of SHs to ensure sufficient video delivery capacity while maintaining low time and memory complexity for the SPA problem. In comparison, the SPA is better than the SSA thanks to optimal power allocation and obviously better than the other AVE, MIN, and MBS.

Finally, we evaluate the capacity performance of all schemes versus the transmission powers from the MBS to the RU (P_{RU}) and to the SH (P_{SH}) changing in the range of [1, 50] W. In Figure 3.6, the SPA is not affected by P_{RU} because P_{RU} is not high enough to make the MBS send all the requested videos to the RU. Similarly, the increase of P_{RU} starts to affect the SSA when it is about 35 W, i.e., the

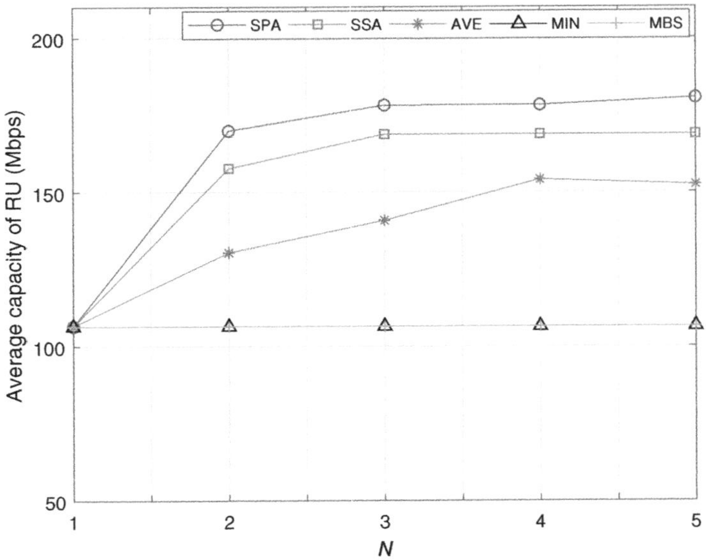

Figure 3.5 Video capacity delivered to RU versus N. *Source*: Reproduced from [1] / IEEE.

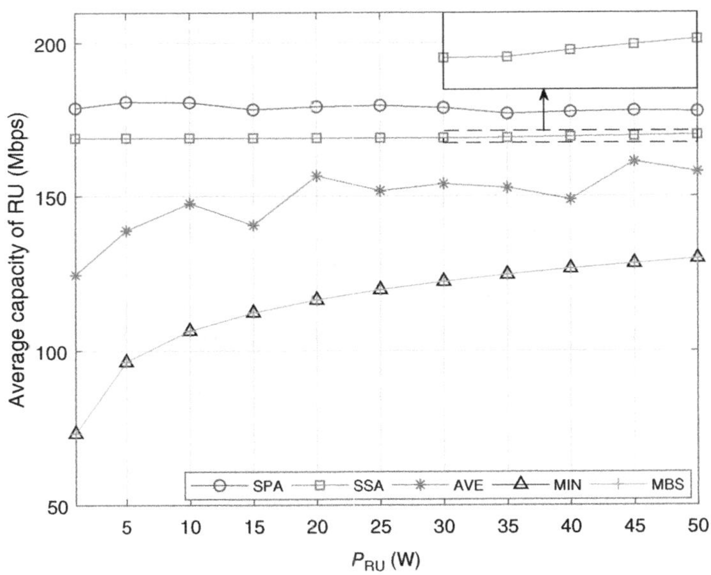

Figure 3.6 System capacity delivered to RU versus P_{RU}. *Source*: Reproduced from [1] / IEEE.

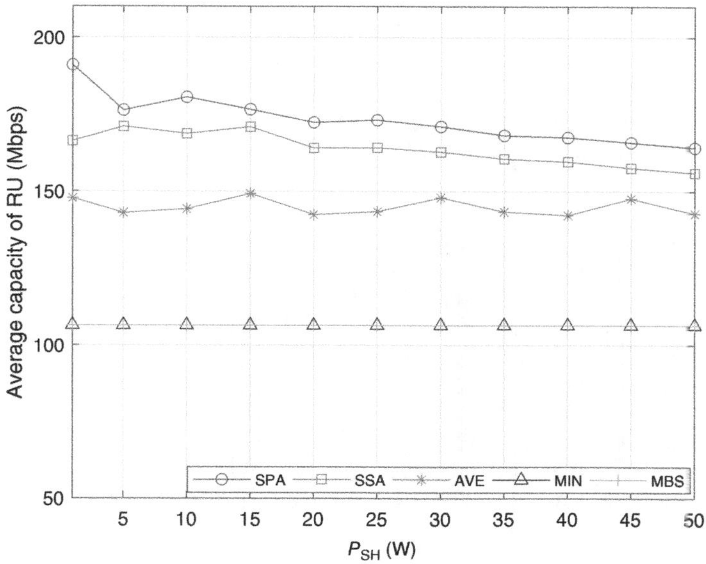

Figure 3.7 System capacity delivered to RU versus P_{SH}. *Source*: Reproduced from [1] / IEEE.

MBS can collaborate more with the SSA in sending the requested videos to the RU so as to improve the performance. Obviously, the increase of P_{RU} directly makes the performance of the AVE, MIN, and MBS become higher. Meanwhile, as shown in Figure 3.7, although increasing P_{SH} yields higher opportunities for sharing the downlink spectrum resources (3.9e), it significantly decreases the capacity delivered to the RU since the interference impact on the RU increases (3.1). It is clear that the increase of P_{SH} does not affect the performance of the MIN and MBS, but slightly decreases that of the AVE due to the performance reduction of the SPA.

3.5.3.2 Power Consumption Performance

We further investigate the total transmission powers consumed by all the Txs of the SPA and SSA versus α and γ_0 as shown in Figures 3.8 and 3.9. Figure 3.8 indicates that besides providing the highest video capacity delivered to the RU as aforementioned, the SPA consumes low transmission power, even a little bit lower than the SSA does. Figure 3.9 shows that the power consumption reduces if fewer MHMH D2D communications are established. The results also validate that the constraint (3.9g) on total transmission power consumption (P) always holds when solving the SPA problem to optimally allocate the transmission powers to all the Txs for MHMP D2D communications.

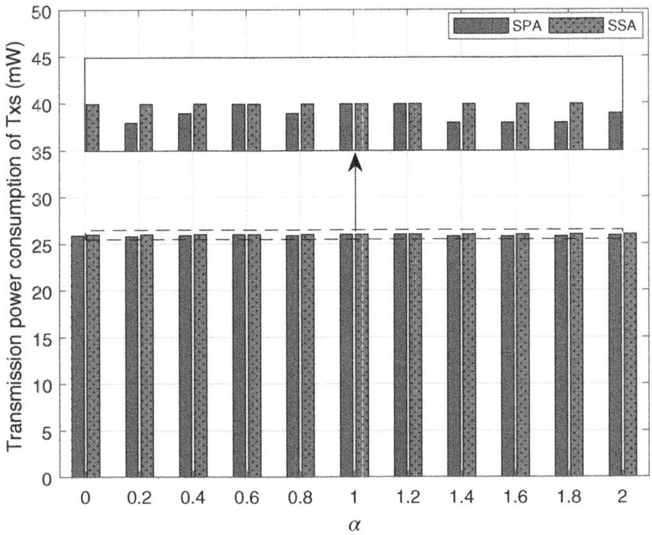

Figure 3.8 Transmission power consumption of Txs versus α. *Source*: Reproduced from [1] / IEEE.

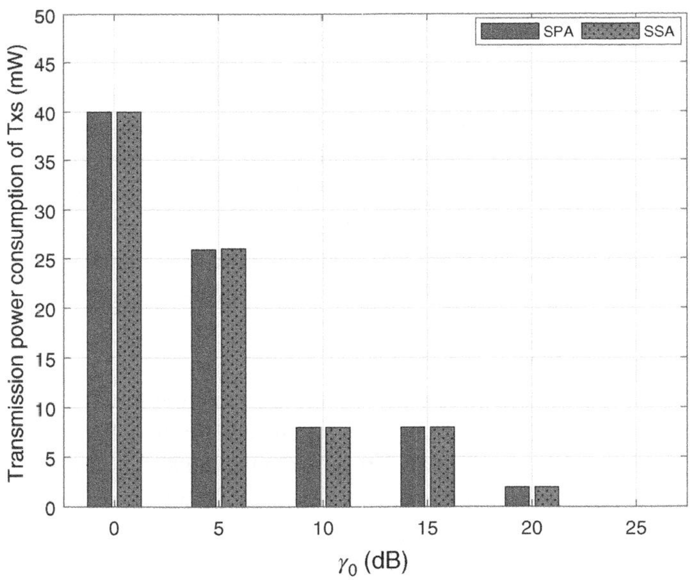

Figure 3.9 Transmission power consumption of Txs versus γ_0. *Source*: Reproduced from [1] / IEEE.

3.6 Conclusion

In this chapter, we have proposed an SPA solution for VAS in 6G networks. The SPA solution exploited not only the storage and cached video resources but also the available downlink spectrum and energy resources of the MU to establish the underlay MHMP D2D communications for offloading the requested videos. The proposed SPA solution enabled the RU to flexibly access the VAS from both the MBS over normal cellular communications and the VH over MHMP D2D communications. To do so, the SPA problem was formulated and solved by GA for the optimal values of (i) downlink spectrum resources shared by the SH, (ii) transmission powers allocated to the Txs of D2D hops, and (iii) transmission mode selection for whether selecting normal cellular communications or MHMP D2D communications to serve the RU. Simulation results demonstrated that GA was feasible to solve the SPA optimization problem, and the SPA solution outperformed the other conventional schemes in terms of providing the highest video delivery capacity while consuming low transmission power. The analyses and discussions with respect to many system parameters and performance metrics provide helpful insights into designing VAS over MHMP D2D communications in 6G networks.

References

1 Q.-N. Tran, N. S. Vo, M.-P. Bui, T.-M. Phan, Q.-A. Nguyen, and T. Q. Duong. Spectrum sharing and power allocation optimised multihop multipath D2D video delivery in beyond 5G networks. *IEEE Trans. Cogn. Commun. and Netw.*, 8(2):919–930. 2022.

2 X. Zhang, Y. Wang, R. Sun, and D. Wang. Clustered device-to-device caching based on file preferences. In *Proc. IEEE 27th Annu. Inter. Symp. Pers., Indoor Mobile Radio Commun.*, pages 1–6, Valencia, Spain, Sep. 2016.

3 N.-S. Vo, T. Q. Duong, T. D. Hoang, and A. Kortun. Optimal video streaming in dense 5G networks with D2D communications. *IEEE Access*, 6:209–223. 2017.

4 J. Song and W. Choi. Mobility-aware content placement for device-to-device caching systems. *IEEE Trans. Wirel. Commun.*, 18(7):3658–3668. 2019.

5 M.-P. Bui, N.-S. Vo, S. Q. Nguyen, and Q.-N. Tran. Social-aware caching and resource sharing maximized video delivery capacity in 5G ultra-dense networks. *Springer Mobile Netw. Appl.*, 25:2037–2049. 2019.

6 N.-S. Vo, M.-P. Bui, P. Q. Truong, C. Yin, and A. Masaracchia. Multi-tier caching and resource sharing for video streaming in 5G ultra-dense networks. *IEEE Commun. Lett.*, 24(7):1500–1504. 2020.

7 M. Rim and C. G. Kang. Cache partitioning and caching strategies for device-to-device caching systems. *IEEE Access*, 9:8192–8211. 2021.

8 Y. Fu, L. Salaün, X. Yang, W. Wen, and T. Q. S. Quek. Caching efficiency maximization for device-to-device communication networks: A recommend to cache approach. *IEEE Trans. Wirel. Commun.*, 20(10):6580–6594. 2021.

9 B. Peng, T. Peng, Z. Liu, Y. Yang, and C. Hu. Cluster-based multicast transmission for device-to-device (D2D) communication. In *Proc. IEEE Veh. Technol. Conf.*, pages 1–5, Las Vegas, NV, Sep. 2013.

10 R. Trestian, Q.-T. Vien, H. X. Nguyen, and O. Gemikonakli. ECO-M: Energy-efficient cluster-oriented multimedia delivery in a LTE D2D environment. In *Proc. IEEE Inter. Commun. Conf.*, pages 1–7, London, UK, June 2015.

11 S.-Z. Jian and T.-C. Hou. Enabling efficient device to device multicast communication by clustering and resource sharing. In *Proc. IEEE Wireless Commun. and Networking Conf.*, pages 1–6, Marrakesh, Morocco, Oct. 2019.

12 N.-S. Vo, T.-M. Phan, M.-P. Bui, X.-K. Dang, V. T. Nguyen, and C. Yin. Social-aware spectrum sharing and caching helper selection strategy optimized multicast video streaming in dense D2D 5G networks. *IEEE Syst. J.*, 15(3):3480–3491. 2021.

13 T.-M. Phan, N.-S. Vo, M.-P. Bui, Q.-N. Tran, H. M. Nguyen, and A. Masaracchia. Downlink resource sharing and multi-tier caching selection maximized multicast video delivery capacity in 5G ultra-dense networks. In *Proc. EAI Inter. Conf. Industrial Netw. and Intell. Syst.*, pages 19–31, Ha Noi, Vietnam, Aug. 2020.

14 Q. Li, X. Wang, D. Wang, and L. Song. Analysis of a cooperative caching-multicast strategy in D2D-aided networks. *IEEE Commun. Lett.*, 10(12):2614–2618. 2021.

15 M. Choi, M. Shin, and J. Kim. Dynamic video delivery using deep reinforcement learning for device-to-device underlaid cache-enabled Internet-of-vehicle networks. *J. Commun. Netw.*, 23(2):117–128. 2021.

16 Q. Xia, Z. Jiao, and Z. Xu. Online learning algorithms for context-aware video caching in D2D edge networks. *IEEE Trans. on Parallel and Distrib. Syst.*, 35(1):1–19. 2024.

17 B. Ying and A. Nayak. A power-efficient and social-aware relay selection method for multi-hop D2D communications. *IEEE Commun. Lett.*, 22(7):1450–1453. 2018.

18 J. Gui and J. Deng. Multi-hop relay-aided underlay D2D communications for improving cellular coverage quality. *IEEE Access*, 8:14318–14338. 2018.

19 X. Lu, J. Zheng, C. Liu, and J. Xiao. A mobility and activeness aware relay selection algorithm for multi-hop D2D communication underlaying cellular networks. In *Proc. IEEE Inter. Commun. Conf.*, pages 1–6, Paris, France, May 2017.

20 H. Yuan, W. Guo, Y. Jin, S. Wang, and M. Ni. Interference-aware multi-hop path selection for device-to-device communications in a cellular interference environment. *IET Commun.*, 11(11):1741–1750. 2017.

21 L. Wei, R. Q. Hu, Y. Qian, and G. Wu. Energy efficiency and spectrum efficiency of multihop device-to-device communications underlaying cellular networks. *IEEE Trans. Veh. Technol.*, 65(1):367–380. 2016.

22 J. Xie, Y. Jia, W. Wen, Z. Chen, and L. Liang. Dynamic D2D multihop offloading in multi-access edge computing from the perspective of learning theory in games. *IEEE Trans. Netw. Service Manag.*, 20(1):305–318. 2023.

23 A. Celik, J. Tetzner, K. Sinha, and J. Matta. 5G device-to-device communication security and multipath routing solutions. *Appl. Netw. Sci.*, 4(102):1–24. 2019.

24 C. Tata and M. Kadoch. Secure multipath routing algorithm for device-to-device communications for public safety over LTE heterogeneous networks. In *Inter. Conf. Future Internet of Things and Cloud*, pages 1–6, Rome, Italy, Oct. 2015.

25 R. E. Ahmed. A low-overhead multi-hop routing protocol for D2D communications in 5G. *J. Commun.*, 16(5):1–7. 2021.

26 E. Arribas and V. Mancuso. Multi-path D2D leads to satisfaction. In *Inter. Symp. A World of Wireless, Mobile and Multimedia Netw.*, pages 1–7, Macau, China, July 2017.

27 E. Arribas and V. Mancuso. Achieving per-flow satisfaction with multi-path D2D. *Ad Hoc Netw.*, 106(1):1–16. 2020.

28 B. Liu, Y. Cao, W. Wang, and T. Jiang. Energy budget aware device-to-device cooperation for mobile videos. In *Proc. IEEE Global Commun. Conf.*, pages 1–7, San Diego, CA, Dec. 2015.

29 Q. Wang, W. Wang, S. Jin, H. Zhu, and N. T. Zhang. Joint coding mode and multi-path selection for video transmission in D2D-underlaid cellular network with shared relays. In *Proc. IEEE Global Commun. Conf.*, pages 1–6, San Diego, CA, Dec. 2015.

30 Q.-N. Tran, N.-S. Vo, M.-P. Bui, V.-C. Phan, Z. Kaleem, and T. Q. Duong. Resource sharing and segment allocation optimized video streaming over multi-hop multi-path in dense D2D 5G networks. In *Proc. EAI Inter. Conf. Industrial Netw. and Intell. Syst.*, pages 26–39, Ho Chi Minh City, Vietnam, Aug. 2019.

31 M.-C. Lee, M. Ji, and A. F. Molisch. Optimal throughput-outage analysis of cache-aided wireless multi-hop D2D networks. *IEEE Trans. Commun.*, 69(4):2489–2504. 2021.

32 A. Bhardwaj and S. Agnihotri. Energy- and spectral-efficiency trade-off for D2D-multicasts in underlay cellular networks. *IEEE Wirel. Commun. Lett.*, 7(4):546–549. 2018.

33 L. Breslau, P. Cao, L. Fan, G. Phillips, and S. Shenker. Web caching and Zipf-like distributions: Evidence and implications. In *Proc. IEEE Inter. Conf. on Computer Commun.*, pages 126–134, New York, NY, Mar. 1999.

34 A. Chipperfield, P. Fleming, H. Pohlheim, and C. Fonseca. *Genetic Algorithm TOOLBOX for Use with Matlab Version 1.2 Users Guide*. University of Sheffield, 1994.

35 T. Fang and L. P. Chau. GOP-based channel rate allocation using genetic algorithm for scalable video streaming over error-prone networks. *IEEE Trans. Image Process.*, 15(6):1323–1330. 2006.

36 Q.-N. Tran, N.-S. Vo, T.-M. Phan, M.-P. Bui, M.-N. Nguyen, and A. Kortun. Downlink resource allocation maximized video delivery capacity over multi-hop multi-path in dense D2D 5G networks. In *Proc. Inter. Conf. Recent Advances on Signal Processing, Telecommun. & Computing*, pages 72–76, Ha Noi, Vietnam, Aug. 2020.

4

Social-aware Spectrum Sharing and Caching Helper Selection for Video Multicasting in D2D Networks

4.1 Introduction

The forthcoming multimedia applications and services (MAS), especially augmented reality (AR) and virtual reality (VR), over 6G networks require an ever-high data transmission rate and massive access.[1] Besides developing new spectrum (e.g., THz bands), integrated network architectures (e.g., low- or very low-earth-orbit satellites, digital twins), and disruptive technologies (sensing and AI) [2], to realize 6G networks, the other characteristics of data, aspects of mobile users (MUs), available system resources, and emerging techniques should be fully utilized. This approach provides the MUs with high quality of service (QoS) while ensuring fairness among the MUs requesting MAS.

In particular, the authors in [3] studied millimeter wave and sub-THz systems assisted by multicasting techniques encompassing reconfigurable intelligent surfaces to meet the increase in MAS data transmission rate while reducing system resource consumption. Multicasting can be combined with scalable encoding and non-orthogonal multiple access (NOMA) to overcome the challenge of huge size, high quality, and low latency of VR videos [4, 5]. To ensure low latency for VR video processing and transmitting prototype systems, both multicasting and edge computing techniques are applied to live VR video streaming in [6]. In [7], a layer-based scalable video caching technique assisted by NOMA solution was deployed in coordination with multicasting and power allocation to maximize the caching-aided data rate.

Multicasting plays an important role in 6G multimedia communications. Typically, in various MASs (such as e-learning, emergency communications, disaster recovery, public safety, over-the-air software updates, news and advertising, AR/VR, and autonomous driving), there exist a number of MUs having the same content interest. Multicasting makes serving multiple MUs simultaneously

1 This chapter has been published partly in [1].

possible, thus reducing the service time and resource consumption (energy, spectrum, and bandwidth) while more MUs can enjoy high QoS.

However, in most studies on multicasting technique with other solutions deployed at the macro base stations (MBS), the available content, energy, and spectrum resources over a much shorter transmission range in dense D2D networks were not considered. Additionally, the unfairness in QoS among the MUs caused by the diverse wireless channels of multicasting has not been addressed. In the studies that applied multicasting technique to D2D networks to improve the system performance, the social attributes of the MUs were not utilized for achieving higher system performance [8–16]. On the other hand, the multicasting technique associated with the MUs' social attributes has been studied, but without utilizing the available spectrum resources of mobile devices for the highest system performance, i.e., maximum system capacity while ensuring the MUs high fairness in QoS [17–21].

In this chapter, we propose a social-aware spectrum sharing and caching helper selection (SSC) optimization method to multicast videos to the MUs in D2D networks. To do so, by exploiting the available cached contents and spectrum resources of mobile devices in the physical layer and social attributes of MUs in the social layer, the clustering, sharing, selecting, and multicasting techniques are synthesized into an SSC optimization problem. The SSC optimization problem is solved by genetic algorithms (GA) for the optimal tripartite set of spectrum helping MUs (SHs), video helping MUs (VHs), and clusters, under the constraint on the fairness in QoS, i.e., capacity fluctuation among the requesting MUs (RUs) and target SINR at the SHs. The optimal solution enables the MBS to establish multiple multicasting sessions to assist itself by offloading the videos to the RUs in different clusters at the highest system capacity while satisfying the QoS fairness for both RUs and SHs. Simulation results are analyzed to demonstrate the feasibility of GA and the benefits of the SSC method compared to other conventional ones.

The rest of this chapter is organized as follows. Section 4.2 introduces a video multicasting system in D2D networks with SSC method. The system is analyzed in Section 4.3, serving as the basis for formulating the SSC optimization problem and proposing the solution in Section 4.4. Section 4.5 presents the performance evaluation of SSC. Finally, Section 4.6 concludes the chapter.

4.2 System Model for Video Multicasting in D2D Networks with SSC

We consider a system of video multicasting in D2D networks with SSC, as shown in Figure 4.1. In the physical layer, one MBS and a number of MUs are connected with each other via cellular, multicasting, and interference links.

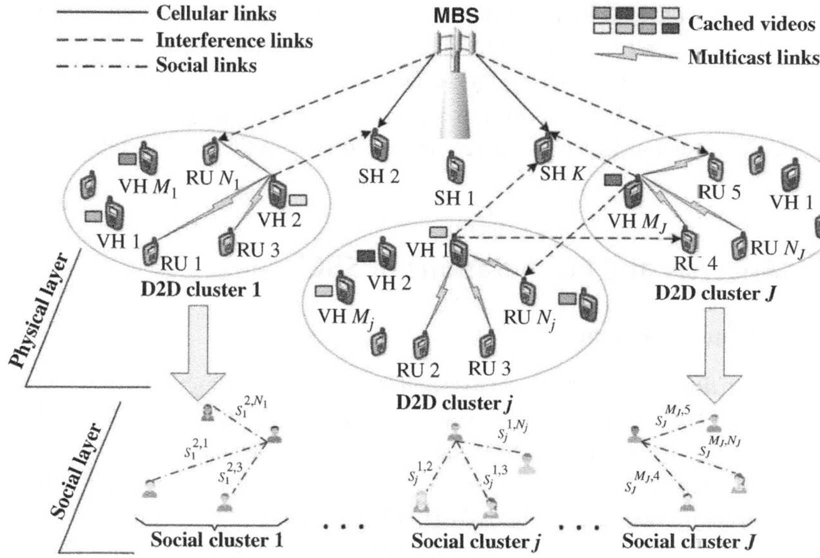

Figure 4.1 Video multicasting in D2D networks with SSC.

The MUs include K SHs for spectrum helping and J clusters in which cluster j has M_j VHs and N_j RUs for video helping and requesting, respectively, $j = 1, 2, ..., J$. Mapped from the physical layer, the social layer provides the social relationships between the VHs and the RUs. Identifying the social relationships makes the SSC method more efficient by selecting the optimal tripartite set of SHs, VHs, and clusters to serve the RUs.

When the MBS detects that numerous RUs are requesting a video, the system operates as follows. First, the MBS collects the system parameters to divide the system into J clusters by using D2D clustering technique [22]. As a result, cluster j has M_j VHs and N_j RUs to expand the coverage area of D2D communications. Then, the MBS formulates the SSC optimization problem for finding the optimal binary index $v_j^{(k,m)}$ representing a tripartite set of SH k, VH m, and cluster j, to maximize the total video capacity multicasted to N_j RUs, $k = 1, 2, ..., K$ and $m = 1, 2, ..., M_j$. Here, $v_j^{(k,m)} = 1$ if SH k agrees to share its spectrum resource with VH m for multicasting the requested video to the RUs in cluster j; otherwise, $v_j^{(k,m)} = 0$.

To enhance efficiency, the SSC optimization problem considers the constraints on the spectrum sharing scheme and the capacity fluctuation amongst the RUs. It also incorporates the interference impact on the SHs when their spectrum resources are reused by the VHs for D2D multicast communications.

Finally, based on the optimal index $\upsilon_j^{(k,m)}$, the MBS assigns the SHs and the VHs to serve the RUs in all clusters by D2D multicast communications. It is noted that if the requested video is not cached in any VHs in a cluster or the multicast capacity delivered from a VH to the RUs in a cluster is lower than that done from the MBS, the RUs normally receive the requested video from the MBS instead of from the VH.

4.3 Analysis and Formulations of SSC

4.3.1 Social Relationship

The social attributes describe how close and frequent the interactions amongst the MUs are, and thus it plays an important role in implementing the SSC method. In this chapter, the Indian Buffet Model (IBM) [23–27] is applied to capture the social relationships between the VHs and the RUs based on the number and the duration of their encounters. The IBM enables us to obtain the probability that a pair of VH and RU in a cluster has a close social relationship to send a video or not. The IBM is detailed in the sequel.

We define $I_j^{(m,n)}$ as the number of encounters of VH m and RU n in the cluster j, $n = 1, 2, \ldots, N_j$. The encounter i, $i = 1, 2, \ldots, I_j^{(m,n)}$, has the duration of $D_{i,j}^{(m,n)}$. The mean of encounter duration of VH m and RU n is given by

$$M_j^{(m,n)} = \frac{\sum_{i=1}^{I_j^{(m,n)}} D_{i,j}^{(m,n)}}{I_j^{(m,n)}}. \tag{4.1}$$

The variance of the encounter duration, which reflects the encounter duration fluctuation between different pairs of VHs and RUs, is expressed as

$$V_j^{(m,n)} = \frac{\sum_{i=1}^{I_j^{(m,n)}} [D_{i,j}^{(m,n)} - M_j^{(m,n)}]^2}{I_j^{(m,n)}}. \tag{4.2}$$

From (4.1) and (4.2), we obtain the Gamma-based encounter duration distribution and the probability density function of the encounter duration as

$$X \sim \Gamma(\kappa_j^{(m,n)}, \theta_j^{(m,n)}) = \Gamma\left(\frac{(M_j^{(m,n)})^2}{V_j^{(m,n)}}, \frac{V_j^{(m,n)}}{M_j^{(m,n)}}\right), \tag{4.3}$$

$$f(x; \kappa_j^{(m,n)}, \theta_j^{(m,n)}) = \frac{1}{(\theta_j^{(m,n)})^{\kappa_j^{(m,n)}} \Gamma(\kappa_j^{(m,n)})} x^{\kappa_j^{(m,n)}-1} e^{-\frac{x}{\theta_j^{(m,n)}}}, \tag{4.4}$$

where $\Gamma(\kappa_j^{(m,n)}) = \int_0^{\infty} t^{\kappa_j^{(m,n)}-1} e^{-t} dt$.

The social-based successful probability to send the video of duration T_j (s) from VH m to RU n in cluster j is

$$S_j^{(m,n)} = 1 - \int_0^{T_j} f(u; \kappa_j^{(m,n)}, \theta_j^{(m,n)}) du = 1 - \frac{\gamma(\kappa_j^{(m,n)}, \frac{T_j}{\theta_j^{(m,n)}})}{\Gamma(\kappa_j^{(m,n)})}, \tag{4.5}$$

where $\gamma(\kappa_j^{(m,n)}, \frac{T_j}{\theta_j^{(m,n)}}) = \int_0^{\frac{T_j}{\theta_j^{(m,n)}}} t^{\kappa_j^{(m,n)}-1} e^{-t} dt$.

We can standardize the social-based successful probability in (4.5) as

$$s_j^{(m,n)} = \frac{S_j^{(m,n)}}{\sum_{l=1}^{N_j} S_j^{(m,l)}}. \tag{4.6}$$

4.3.2 Wireless Channels

Consider a transmitter (Tx) and a receiver (Rx) in cluster j, let $h_j^{\mathrm{Tx,Rx}}$ be the exponential power fading coefficient modeled as an independent and identically exponential distribution with unit mean, and $g_j^{\mathrm{Tx,Rx}} = (d_j^{\mathrm{Tx,Rx}})^{-\eta}$ be the standard power law path loss function with path loss exponent η, the wireless channel gain between Tx and Rx in cluster j is expressed as [28, 29]

$$G_j^{\mathrm{Tx,Rx}} = h_j^{\mathrm{Tx,Rx}} g_j^{\mathrm{Tx,Rx}}, \tag{4.7}$$

where Tx $\in$ {MBS represented by 0, VH m}, Rx $\in$ {SH k, RU n}, and $d_j^{\mathrm{Tx,Rx}}$ is the distance from Tx to Rx in cluster j.

4.3.3 Capacity at RUs

In SSC strategy, the video capacity at the RUs comes from the MBS or the VHs. To derive the capacity, we first compute the SINR at RU n from VH m in cluster j obtained by reusing the downlink resource of SH k, given by

$$\gamma_j^{(k,m,n)} = \frac{\upsilon_j^{(k,m)} s_j^{(m,n)} P_j^{(m)} G_j^{(m,n)}}{\sigma_0 + P^{(0,k)} G_j^{(0,n)} + I_j^{(k,n)}}, \tag{4.8}$$

where $P_j^{(m)}$ is the multicast D2D transmission power of VH m in cluster j, $P^{(0,k)}$ is the transmission power of MBS to SH k, σ_0 is the power of additive white Gaussian noise, and $I_j^{(k,n)}$, which is the interference impact on RU n in cluster j transmitted by the VHs in other clusters that reuse the same downlink resource shared by SH k, is expressed as

$$I_j^{(k,n)} = \sum_{j'=1,j'\neq j}^{J} \sum_{m'=1}^{M_{j'}} v_{j'}^{(k,m')} s_{j'}^{(m',n)} P_{j'}^{(m')} G_{j'}^{(m',n)}. \tag{4.9}$$

If the transmission from VH m to RU n cannot be established due to poor capacity or/and unfairness in QoS, RU n in cluster j is normally served by the MBS. The signal-to-noise ratio (SNR) of the channel between the MBS and RU n is expressed as

$$\gamma_j^{(0,n)} = \frac{\prod_{k=1}^{K} \prod_{m=1}^{M_j} \left(1 - v_j^{(k,m)} s_j^{(m,n)}\right) P_j^{(0,n)} G_j^{(0,n)}}{\sigma_0}, \tag{4.10}$$

where $P_j^{(0,n)}$ is the transmission power of MBS to RU n in the cluster j.

Finally, the total video capacity at the RUs delivered from the MBS and multicasted from the VHs is given by

$$C = W \sum_{j=1}^{J} \left[\sum_{k=1}^{K} \sum_{m=1}^{M_j} \sum_{n=1}^{N_j} C_j^{(k,m,n)} + \sum_{n=1}^{N_j} C_j^{(0,n)} \right], \tag{4.11}$$

where

$$C_j^{(k,m,n)} = \log_2(1 + \gamma_j^{(k,m,n)}), \tag{4.12}$$

$$C_j^{(0,n)} = \log_2(1 + \gamma_j^{(0,n)}), \tag{4.13}$$

and W is the system bandwidth.

4.3.4 Capacity Fluctuation Amongst RUs

In SSC method, the capacity values received at the RUs might be different from each other due to the diverse characteristics of wireless channels of multicasting. The capacity fluctuation introduces the unfairness in QoS amongst the RUs. To limit the fluctuation, the standard deviation of capacity, which must be considered as a constraint in the SSC optimization problem, is computed as

$$\sigma_C = \sqrt{\frac{\sum_{j=1}^{J} \left\{ \sum_{n=1}^{N_j} \left[W \left(\sum_{k=1}^{K} \sum_{m=1}^{M_j} C_j^{(k,m,n)} + C_j^{(0,n)} \right) - \overline{C} \right]^2 \right\}}{\sum_{j=1}^{J} N_j}}, \tag{4.14}$$

where $\overline{C} = \dfrac{C}{\sum_{j=1}^{J} N_j}$.

4.3.5 SINR at SHs

The unfairness in QoS happens not only to the RUs but also to the SHs when they share downlink resources with the VHs for multicasting. When SH k shares its downlink resource with the VHs in multiple clusters, its SINR is reduced due to the interference generated by the VHs and thus it is unfair to the SHs. To limit the interference impact on the SHs, the SINR at SH k is computed as

$$\gamma^{(k)} = \frac{P^{(0,k)}G^{(0,k)}}{\sigma_0 + \sum_{j=1}^{J}\sum_{m=1}^{M_j} v_j^{(k,m)} P_j^{(m)} G_j^{(m,k)}}. \tag{4.15}$$

where $G^{(0,k)}$ is the channel gain from the MBS to SH k similarly given in (4.7).

4.4 SSC Optimization Design

4.4.1 SSC Optimization Problem

The SSC optimization problem aims at maximizing the total video capacity at the RUs given in (4.11) by finding the optimal index of the downlink resource sharing and caching helper selection $v_j^{(k,m)}$ in each cluster. Together with the constraints on spectrum sharing scheme and the fairness in QoS of the RUs and the SHs, the SSC optimization problem is formulated as

$$\max_{v_j^{(k,m)}} C, \tag{4.16a}$$

$$\text{s.t.} \begin{cases} \sum_{k=1}^{K}\sum_{m=1}^{M_j} v_j^{(k,m)} \leq 1, \forall j = 1, 2, ..., J, \\ \sigma_C \leq \sigma, \\ \sum_{j=1}^{J}\sum_{m=1}^{M_j} v_j^{(k,m)} P_j^{(m)} G_j^{(m,k)} \leq \dfrac{P^{(0,k)}G^{(0,k)}}{\gamma_0} - \sigma_0, \\ k = 1, 2, ..., K, \end{cases} \tag{4.16b}$$

where in (4.16b), the first constraint is to efficiently exploit the downlink resources of the SHs such that each SH can share with up to J VHs in J clusters, but with up to one VH in each cluster. For the fairness in QoS of the RUs and the SHs, the second constraint is to limit the capacity fluctuation among the RUs by σ and the third constraint comes from (4.15) by letting $\gamma^{(k)} \geq \gamma_0$ to guarantee that the SINR of SH k is not less than the target SINR γ_0 when it shares the downlink resource for D2D multicast communications.

The SSC optimization problem can be solved for the exact optimal results by applying the exhausted search algorithm (ESA) [30, 31]. The exact optimal results consist of J optimization binary matrices, i.e., $V^*_{K \times M_1}$, $V^*_{K \times M_2}$,..., $V^*_{K \times M_J}$, selected from the following search space

$$\begin{aligned}\mathcal{V} = \{&V^1_{K \times M_1}, V^2_{K \times M_1}, \dots, V^{2^{KM_1}}_{K \times M_1}; V^1_{K \times M_2}, V^2_{K \times M_2}, \dots, \\ &V^{2^{KM_2}}_{K \times M_2}; \dots; V^1_{K \times M_J}, V^2_{K \times M_J}, \dots, V^{2^{KM_J}}_{K \times M_J}\},\end{aligned} \tag{4.17}$$

where $V^{\upsilon}_{K \times M_j}$, $\upsilon = 1, 2, \dots, 2^{KM_j}$, is the $K \times M_j$ binary matrix.

However, if J, K, and M_j increase, the ESA becomes impractical in a large-scale dense D2D networks due to high computational cost, i.e., the memory and time complexity is of $\mathcal{O}(2^{K \times \sum_{j=1}^{J} M_j})$. In practice, an approximate optimal result obtained at low computational cost is more feasible than an exact optimal result done at high computational cost. In this chapter, GAs, which are the stochastic global search solution based on the selection of natural biological evolution [32], is applied to solve the SSC optimization problem for exact or approximate optimal result at reasonable computational cost. The GA solution is detailed in the following sections.

4.4.2 SSC Optimization Solution with GA

GA is implemented based on a randomly initial population of individuals, i.e., potential solutions, that get better approximations to a global optimal solution after each generation following the principles of evolutionary natural selection. The substantial benefit of GA is that it is able to search many optimal points in parallel, i.e., the search space is not required to be unimodal [33, 34]. GA can make use of random processing to avoid being trapped in the local optima. In addition, GA is flexible since it can provide a number of potential exact and approximate solutions for a given optimization problem. The final proper solution corresponding to a particular computational resource requirement is defined by system designers [34].

While the memory and time complexity of ESA exponentially increase with respect to J, K, and M_j, GA conserves a memory complexity of $O(N_P \times N_B)$, where N_P is the population size, i.e., the number of individuals (or potential solutions), and $N_B = K \times \sum_{j=1}^{J} M_j$ is the number of bits used to represent an individual. The time complexity of GA depends on the values of N_P and N_B, the operators of GA (selection, crossover, and mutation), the fitness function evaluation, and the convergence criteria defined by system designers [28, 34].

GA is able to solve an optimization problem with simple constraints of lower and upper bounds but not with the complicated ones. This issue can be resolved by using a penalty function [33] to convert the SSC optimization with complicated constraints to an unconstrained optimization problem. To do so, the complicated constraints in (4.16b) are rewritten as

$$\begin{cases} \Delta V_j = 1 - \sum_{k=1}^{K} \sum_{m=1}^{M_j} v_j^{(k,m)} \geq 0, j = 1, 2, \ldots, J, \\ \Delta\sigma = \sigma - \sigma_C \geq 0. \\ \Delta\gamma^{(k)} = \dfrac{P^{(0,k)} G^{(0,k)}}{\gamma_0} - \sigma_0 - \sum_{j=1}^{J} \sum_{m=1}^{M_j} v_j^{(k,m)} P_j^{(m)} G_j^{(m,k)} \geq 0, \\ k = 1, 2, \ldots, K, \end{cases} \tag{4.18}$$

Then, we obtain the penalty function as

$$F = \lambda_1 \sum_{j=1}^{J} \left(\min\{0, \Delta V_j\}\right)^2 + \lambda_2 (\min\{0, \Delta\sigma\})^2 + \\ \lambda_3 \sum_{k=1}^{K} (\min\{0, \Delta\gamma^k\})^2, \tag{4.19}$$

where λ_1, λ_2, and λ_3 are the constraint violation degrees used to punish the individuals which violate the constraints. The individuals are punished by reducing C by F so that they become better, i.e., satisfying the constraints, after each generation. Finally, we apply GA to solve the following unconstrained optimization problem

$$\max_{v_j^{(k,m)}} C_F = C - F. \tag{4.20}$$

In this chapter, the GA, which is implemented for the SSC optimization problem given in [34], is presented in Algorithm 4.1. Given the number of generations N_G, GA terminates if one of the following conditions (TC) holds:

- The error of $\overline{C}_F$ between two generations (Gen − 1 and Gen) and the penalty F are less than 10^{-3} in 10 consecutive generations.
- $Gen = N_G$.

Algorithm 4.1 GA for SSC Optimization Problem

Input: Parameters of system in Table 2.2
  N_{P}: Population size (10,000 individuals)
  P_{G}: Generation gap (0.9)
  N_{B}: $K \times \sum_{j=1}^{J} M_j = 150$ bits, i.e., number of bits for each index $\upsilon_j^{(k,m)}$
  P_{C}: Crossover probability (0.9)
  P_{M}: Mutation probability (10^{-8})
  N_{G}: Number of generations (100)
  TC: Termination condition
  $Gen = 1$: Generation count
Output: $\overline{C}_F^{*}(\upsilon_j^{(k,m)*})$

Generate N_{P} random strings (chromosomes), each has $N_{\text{B}} = K \times \sum_{j=1}^{J} M_j$ bits, namely $\{X^{(z)}\}$, to represent individual z, $z = 1, 2, \ldots, N_{\text{P}}$
Compute (4.20) to have N_{P} fitness values $\{\overline{C}_F(X^{(z)})\}$
while TC does not hold **do**
  Put $\{X^{(z)}\}$ associated with $\{\overline{C}_F(X^{(z)})\}$ into the mating pool for ranking
  Select $N_{\text{PG}} = N_{\text{P}} \times P_{\text{G}}$ individuals with higher fitness values by using stochastic universal sampling operator for breeding [34], to obtain $\{X^{(t)}\}$, $t = 1, 2, \ldots, N_{\text{PG}}$
  Choose a pair of parents in $\{X^{(t)}\}$ to produce the offsprings by using single point crossover with probability P_{C}
  Mutate the offsprings with probability P_{M} to obtain $\{X'^{(t)}\}$ so that the positive genetic features probably lost in the previous steps can be recovered
  Compute (4.20) to have N_{PG} fitness values $\{\overline{C}_F(X'^{(t)})\}$, then reinsert $\{X'^{(t)}\}$ and $\{\overline{C}_F(X'^{(t)})\}$ into the current generation to get $\{X^{(z)}\}$ and $\{\overline{C}_F(X^{(z)})\}$
  $Gen = Gen + 1$
end while
Find the highest fitness value $\overline{C}_F^{*}(\upsilon_j^{(k,m)*}) \in \{\overline{C}_F(X^{(z)})\}$ in the last generation

Source: [35] / IEEE.

4.5 Performance Evaluation of SSC

4.5.1 System Setting

The system parameters are listed in Table 4.1. In addition, since the value of the penalty function F is much smaller than the value of the objective function C, the penalty value is not high enough to punish the individuals to satisfy the constraints. Therefore, the violation degrees are set relatively high, i.e., $\lambda_1 = 10^6$, $\lambda_2 = 10^6$, and $\lambda_3 = 10^{32}$. The way to choose the violation degrees is presented in [28].

4.5.2 GA Convergence Evaluation

As we can see in Figure 4.2, GA quickly converges after 20 generations. In the convergence situation, the mean of fitness values of all individuals is equal to the best fitness value, and the penalty value F is zero. In other words, all the individuals

Table 4.1 Parameters setting.

Symbols	Values
J	5 clusters
K	5 SHs
$\{M_j\}$	$\{2, 4, 6, 8, 10\}$ VHs
$\{N_j\}$	$\{5, 10, 15, 20, 25\}$ RUs
W	5 Mhz
σ_0	10^{-13} W
γ_0	10 dB
σ	25
$P^{(0,k)}, P_j^{(0,n)}$	Fixed to 5W
$P_j^{(m)}$	Randomly uniform distributed in $[0.01, 0.1]$ W
η	4
$\{T_j\}$	$\{20, 10, 15, 13, 5\}$ s
$\kappa_j^{(m,n)}$	Randomly uniform distributed in $[1, 10]$
$\theta_j^{(m,n)}$	Randomly uniform distributed in $[1, 30]$
Distances	**Specifications**
MBS to RUs ($d_j^{(0,n)}$)	Randomly uniform distributed in $[100, 1{,}000]$ m
MBS to SHs ($d_j^{(0,k)}$)	Randomly uniform distributed in $[100, 500]$ m
VHs to RUs	Randomly uniform distributed in $[1, 50]$ m and $[1, 300]$ m for $d_j^{(m,n)}$ and $d_{j'}^{(m',n)}$, respectively
VHs to SHs ($d_j^{(m,k)}$)	Randomly uniform distributed in $[50, 150]$ m

have the same best quality without punishment for violating the constraints. The quick convergence rate demonstrates that the GA is feasible for solving the SSC optimization problem.

We further investigate the performance of GA by taking into account the time complexity and the accuracy versus J and N_P as shown in Tables 4.2 and 4.3. The results in Table 4.2 indicate that if the number of clusters is low ($J = 1$), GA require a small population size ($N_P = 100$) to get converged within a very short time of 0.27 s. When J increases, a larger population size is required to make GA converged within a higher time complexity, but with reasonable value of 930.50 s, at $J = 5$ and $N_P = 10{,}000$. Meanwhile, the small population size ($N_P = 100$) cannot ensure GA converged, i.e., N/A in Table 4.2. Similarly, in Table 4.3, increasing J requires larger population sizes to achieve the highest accuracy. Taking $J = 5$ as an example, $N_P = 500$ and $N_P = 1{,}000$ can make GA converged, but cannot hit 100% accuracy, i.e., only 90.09 and 96.32%, respectively. The highest 100% accuracy

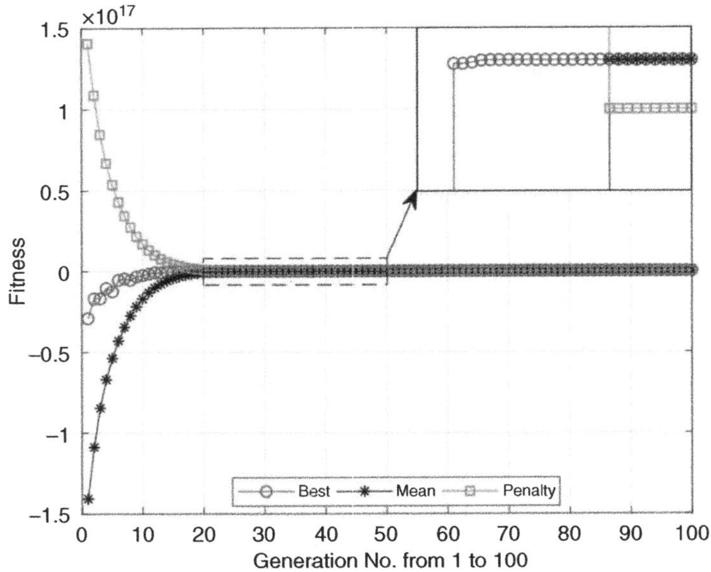

Figure 4.2 Convergence rate of GA. *Source*: Reproduced from [1] / IEEE.

Table 4.2 GAs convergence time (s) versus J and N_P.

J \ N_P	100	500	1,000	5,000	10,000
1	0.27	0.40	0.66	4.61	17.67
3	N/A	5.76	13.00	65.78	158.99
5	N/A	44.55	113.53	486.83	930.50

Table 4.3 GAs accuracy (%) versus J and N_P.

J \ N_P	100	500	1,000	5,000	10,000
1	100	100	100	100	100
3	N/A	100	100	100	100
5	N/A	90.09	96.32	100	100

holds at $N_P = 5{,}000$ and $N_P = 10{,}000$, and thus we select $N_P = 10{,}000$ to ensure both the convergence and accuracy of GA. It is noted that the accuracy is obtained based on the exact optimal results by using the conventional ESA [31] to solve the SSC optimization problem.

4.5.3 SSC Evaluation

To evaluate the performance of SSC, we compare it to other schemes, including nonsocial-aware spectrum sharing and caching helper selection (NSA), only MBS (OMB), average performance (AVE), and minimum performance (MIN). In NSA, the social relationships between the VHs and the RUs are the same. In OMB, the videos are multicasted only from the MBS to the RUs in all clusters. In AVE, we find a set of feasible individuals (solutions) that meet the constraints and compute C on average with respect to each individual. The set of feasible individuals is found by deploying GA to minimize the penalty function F. In MIN, we deploy GAs to minimize $(C + F)$ to obtain the worst case of SSC instead of maximizing C_F. The performance evaluations of SSC, NSA, OMB, AVE, and MIN in terms of system capacity and standard deviation of capacity are presented in the following.

Figure 4.3 plots the system capacity versus the target SINR of SHs γ_0. It is clear that if γ_0 is low, i.e., the SHs agree to degrade their QoS to share their spectrum resources with more VHs for D2D multicast communications, the system capacity is high. In contrast, increasing γ_0 makes less spectrum resource sharing to guarantee the SHs higher QoS, thereby decreasing the system capacity. Obviously, if

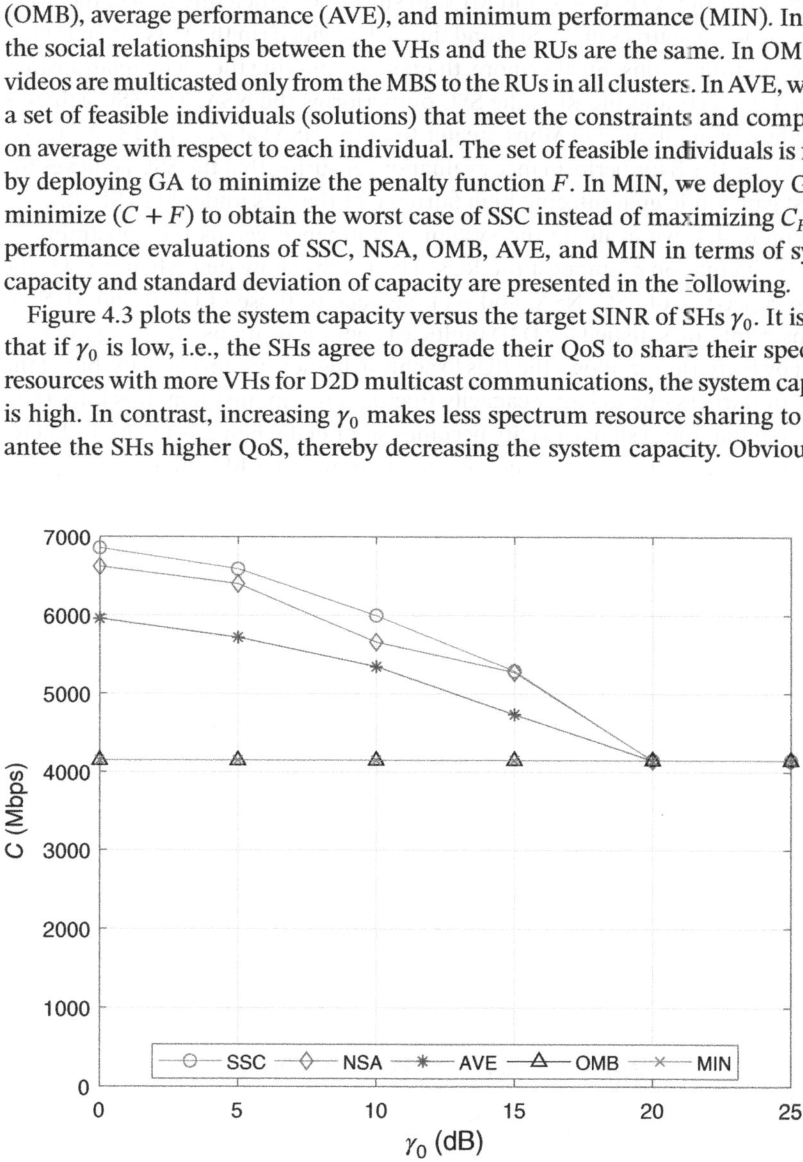

Figure 4.3 System capacity versus the target SINR of SHs γ_0. *Source*: Reproduced from [1] / IEEE.

γ_0 is too high ($\gamma_0 \geq 20$ dB), there is no spectrum resource sharing nor D2D multicast communications, leading to the fact that all the RUs are served by the MBS at the lowest system capacity. In comparison, the SSC and NSA gain higher system capacity than the AVE, OMB, and MIN do since they efficiently utilize the available spectrum resources of the SHs and the cached videos in the VHs for D2D multicast communications. Furthermore, thanks to exploiting the social relationships between the VHs and the RUs, the SSC outperforms the NSA. The SSC achieves the highest gain (about 340 Mbps greater than the NSA) at $\gamma_0 = 10$ dB. We therefore set $\gamma_0 = 10$ dB for performance evaluation, ensuring that the SSC achieves the highest gain while guaranteeing high fairness for the SHs under high SINR.

In Figure 4.4, we evaluate the system performance versus the constraint on capacity fluctuation σ amongst the RUs. The results show that when $\sigma \leq 15$, the system capacities of SSC, NSA, and AVE degrade to those of OMB and MIN, as no constraints are satisfied for D2D multicast communications. Relaxing this constraint by increasing σ allows the RUs to achieve higher system capacity, but at the cost of unfairness due to larger capacity fluctuations among them. It is important to observe that the system capacity becomes saturated when σ is too high. Therefore, we carefully select the value of σ to guarantee the RUs high capacity and

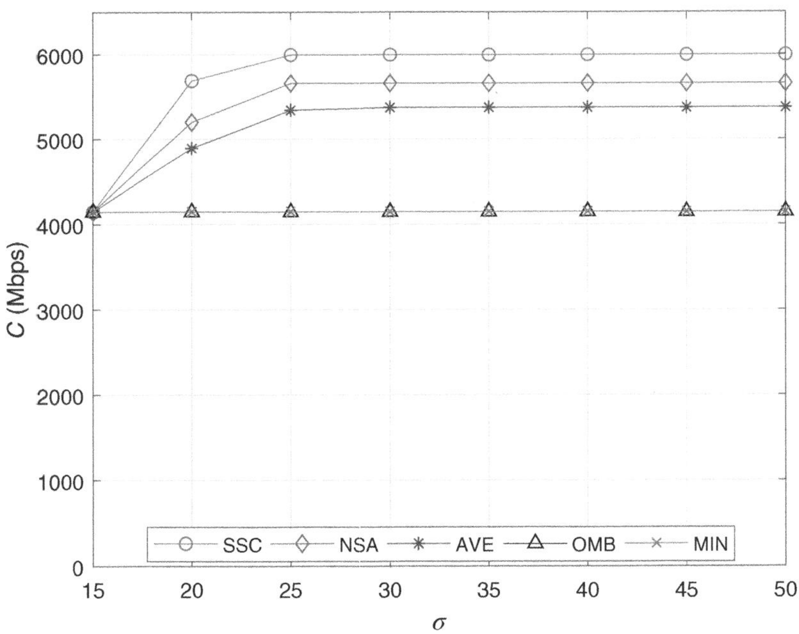

Figure 4.4 System capacity versus the standard deviation of capacity σ.
Source: Reproduced from [1] / IEEE.

fairness, $\sigma = 25$ at the knee of the SSC curve, for example. In comparison, the SSC gains the highest system capacity compared to the NSA, AVE, OMB, and MIN.

It is certain that in Figure 4.5, increasing the number of clusters J introduces higher capacity to serve more RUs. The SSC is only better than the NSA until the number of clusters is high enough ($J = 3$). The reason is that if J is low, the number of VHs and RUs is low too. Therefore, the social relationship factor does not make a significant impact on the system performance resulting in the fact that the SSC cannot exploit the social relationships to outperform the NSA. In other words, the social relationship factor can be efficiently exploited in dense D2D networks to improve the system capacity. In comparison, the SSC and NSA are better than the AVE and significantly surpass the two worst cases of OMB and MIN.

Figure 4.6 illustrates the system capacity versus the number of SHs K. It is easy to see that the more SHs the system has, the better chances the system gets to assign the proper SHs to share their downlink spectrum resources with the VHs. It means that the system capacity is likely to become higher when increasing K. Importantly, the system capacity gets saturated if the new SHs added, i.e., $K = 2, 3, 4, 5$, to the system are not better than the existing SHs. The SSC is always better than the NSA, AVE, OMB, and MIN, except for $K = 0$ without any

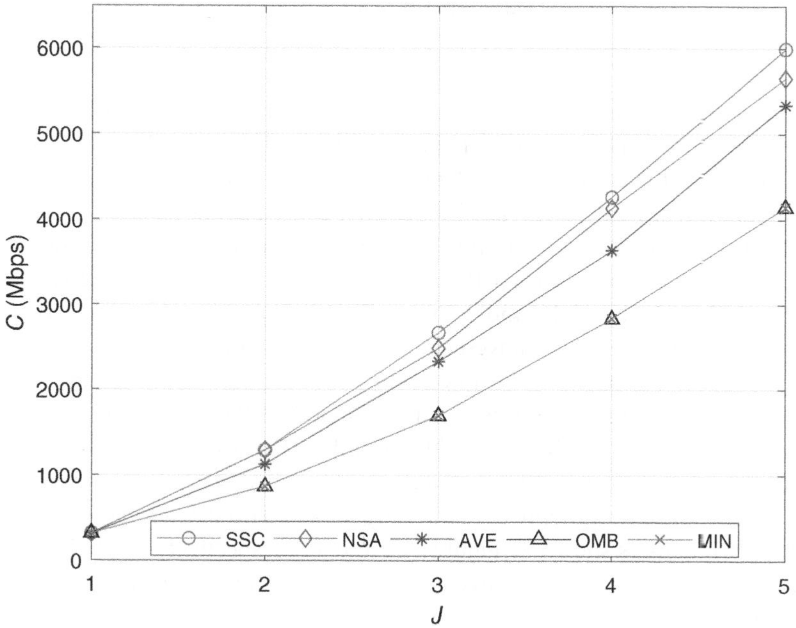

Figure 4.5 System capacity versus the number of clusters J. *Source*: Reproduced from [1] / IEEE.

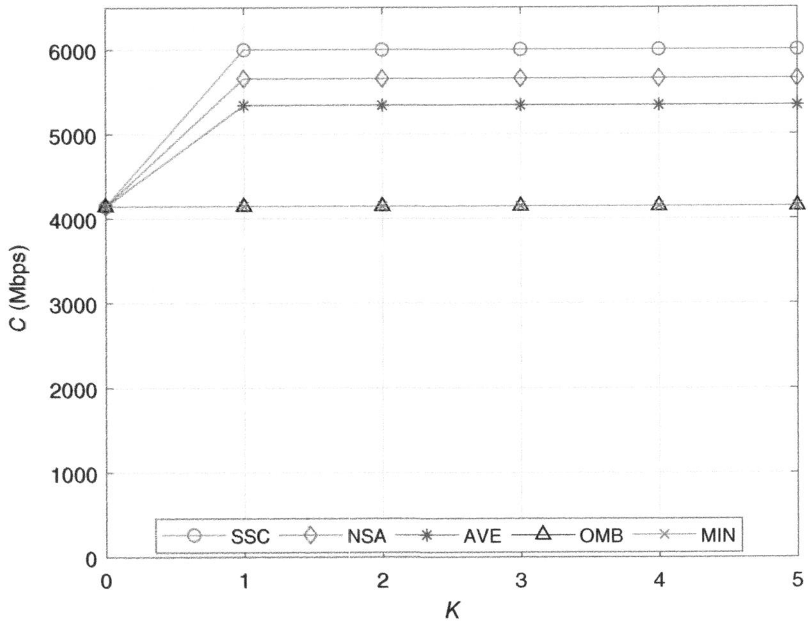

Figure 4.6 System capacity versus the number of SHs K. *Source*: Reproduced from [1] / IEEE.

downlink resource sharing nor D2D multicast communications. It is noted that the saturation does not hold if we keep the number of SHs increased until the new one is better than the existing ones. The better SH is the one located closer to the MBS but far away from the VHs (less interference caused by the VHs).

Finally, we investigate the capacity fluctuation amongst the RUs represented by the standard deviation of capacity σ_C as depicted in Figure 4.7. The results show that although the OMB and MIN provide the lowest system capacity as aforementioned, they yield the lowest standard deviation of capacity to serve the RUs in the highest fairness, i.e., the lowest capacity fluctuation amongst the RUs. The trade-off of the proposed SSC is that it offers the RUs the highest capacity but low fairness. However, the SSC can flexibly control the fairness by constraining the standard deviation of capacity to a reasonable value, i.e., $\sigma = 25$, that gains much higher system capacity and only two times higher standard deviation of capacity than the OMB does as shown in Figures 4.4 and 4.7, respectively. Explicitly, we provide a summary fairness comparison versus the two fairness constraints on γ_0 and σ as presented in Table 4.4. We categorize the fairness into three types including "Excellent," "Good," and "Fair."

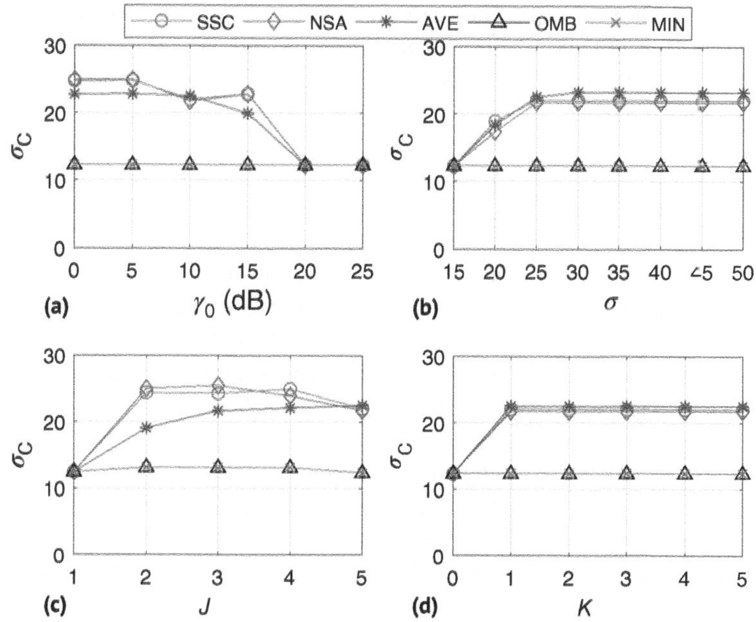

Figure 4.7 Standard deviation of capacity at the RUs. *Source*: Reproduced from [1] / IEEE.

Table 4.4 Fairness comparison.

Constraint \ Scheme	SSC	NSA	AVE	OMB	MIN
γ_0	Fair	Fair	Good	Excellent	Excellent
σ	Good	Good	Fair	Excellent	Excellent

4.6 Conclusion

In this chapter, we have proposed the SSC method to multicast the requested videos to the RUs in D2D networks. The SSC not only applies the emerging caching, D2D communications, and multicasting techniques assisted by downlink spectrum resource sharing solution, but also exploits the social attributes of the VHs and the RUs. The SSC optimization problem was formulated for the optimal tripartite set of SHs, VHs, and clusters under the fairness constraints on the capacity fluctuation amongst the RUs and the target SINR at the SHs. The optimal solution enabled the SSC method to flexibly control the capacity traffic delivered to the RUs from whether the MBS or the VHs over D2D multicast communications to provide the RUs with maximum video capacity while guaranteeing the RUs and the SHs high fairness of QoS. We investigated the pros and cons of the proposed SSC method through the simulation results to demonstrate its efficiency in comparison with other schemes.

References

1 N.-S. Vo, T.-M. Phan, M.-P. Bui, X.-K. Dang, V. T. Nguyen, and C. Yin. Social-aware spectrum sharing and caching helper selection strategy optimized multicast video streaming in dense D2D 5G networks. *IEEE Syst. J.*, 15(3):3480–3491. 2021.

2 N.-S. Vo, A. Masaracchia, Z. Sheng, and T. T. Nguyen. Towards 6G networks: Technologies, services and applications. *Springer Mobile Netw. Appl.*, 28:1142–1144. 2023.

3 N. Chukhno, O. Chukhno, D. Moltchanov, S. Pizzi, A. Gaydamaka, and A. Samuylov. Models, methods, and solutions for multicasting in 5G/6G mmWave and sub-THz systems. *IEEE Commun. Surv. Tutor.*, 26(1):119–159. 2024.

4 N. Gao, G. Liu, M. Feng, X. Hua, and T. Jiang. Non-orthogonal multiple access enhanced scalable 360-degree video multicast. *IEEE Trans. Multimed.*, 26:8488–8503. 2024.

5 N. Gao, Y. Wang, X. Hua, T. Bi, and T. Jiang. 360° video multicast scheduling optimization in orthogonal and nonorthogonal scenarios. *IEEE Internet of Things J.*, 11(10):17128–17140. 2024.

6 N. Gao, J. Zhou, G. Wan, X. Hua, T. Bi, and T. Jiang. Low-latency VR video processing-transmitting system based on edge computing. *IEEE Trans. Broadcast.*, 70(3):862–871. 2024.

7 J. Ma, L. Liu, B. Shang, S. Jere, and P. Fan. Performance analysis and optimization for layer-based scalable video caching in 6G networks. *IEEE/ACM Trans. Netw.*, 31(4):1494–1506. 2023.

8 S. Joshi and R. K. Mallik. Coverage and interference in D2D networks with Poisson cluster process. *IEEE Commun. Lett.*, 22(5):1098–1101. 2018.

9 A. Masaracchia, L. D. Nguyen, T. Q. Duong, and M.-N. Nguyen. An energy-efficient clustering and routing framework for disaster relief network. *IEEE Access*, 7:56520–56532. 2019.

10 R. Amer, M. M. Butt, M. Bennis, and N. Marchetti. Inter-cluster cooperation for wireless D2D caching networks. *IEEE Trans. Wirel. Commun.*, 17(9):6108–6121. 2018.

11 M. Afshang, H. S. Dhillon, and P. H. J. Chong. Fundamentals of cluster-centric content placement in cache-enabled device-to-device networks. *IEEE Trans. Commun.*, 64(6):2511–2526. 2016.

12 Z. Chang, Y. Hu, Y. Chen, and B. Zeng. Cluster-oriented device-to-device multimedia communications: Joint power, bandwidth, and link selection optimization. *IEEE Trans. Veh. Technol.*, 67(2):1570–1581. 2018.

13 H. Ding, X. Wang, D. B. da Costa, and J. Ge. Interference modeling in clustered device-to-device networks with uniform transmitter selection. *IEEE Trans. Wirel. Commun.*, 16(12):7906–7918. 2017.

14 X. Lin, R. Ratasuk, A. Ghosh, and J. G. Andrews. Modeling, analysis, and optimization of multicast device-to-device transmissions. *IEEE Trans. Wirel. Commun.*, 13(8):4346–4359. 2014.

15 L. Militano, M. Condoluci, G. Araniti, A. Molinaro, A. Iera, and G.-M. Muntean. Single frequency-based device-to-device-enhanced video delivery for evolved multimedia broadcast and multicast services. *IEEE Trans. Broadcast.*, 61(2):263–278. 2015.

16 J.-H. Kim, J. Joung, and J. W. Lee. Resource allocation for multiple device-to-device cluster multicast communications underlay cellular networks. *IEEE Commun. Lett.*, 22(2):412–415. 2018.

17 Y. Cao, T. Jiang, X. Chen, and J. Zhang. Social-aware video multicast based on device-to-device communications. *IEEE Trans. Mob. Comput.*, 15(6):1528–1539. 2016.

18 P. Zhao, L. Feng, P. Yu, W. Li, and X. Qiu. A social-aware resource allocation for 5G device-to-device multicast communication. *IEEE Access*, 5:15717–15730. 2017.

19 Y. Wu, D. Wu, L. Yang, X. Shi, L. Ao, and Q. Fu. Matching-coalition based cluster formation for D2D multicast content sharing. *IEEE Access*, 7:73913–73928. 2019.

20 L. Yang, D.Wu, S. Xu, G. Zhang, and Y. Cai. Social-energy-aware user clustering for content sharing based on D2D multicast communications. *IEEE Access*, 6:36092–36104. 2018.

21 L. Feng, P. Zhao, F. Zhou, M. Yin, P. Yu, W. Li, and X. Qiu. Resource allocation for 5G D2D multicast content sharing in social-aware cellular networks. *IEEE Commun. Mag.*, 56(3):112–118. 2018.

22 S. Gyawali, S. Xu, F. Ye, R. Q. Hu, and Y. Qian. A D2D based clustering scheme for public safety communications. In *Proc. IEEE Veh. Technol. Conf.*, pages 1–5, Porto, Portugal, June 2018.

23 X. Zhang, Y. Li, Y. Zhang, J. Zhang, H. Li, S. Wang, and D. Wang. Information caching strategy for cyber social computing based wireless networks. *IEEE Trans. Emerg. Top. Comput.*, 5(3):391–402. 2017.

24 Y. Zhang, E. Pan, L. Song, W. Saad, Z. Dawy, and Z. Han. Social network aware device-to-device communication in wireless networks. *IEEE Trans. Wirel. Commun.*, 14(1):177–190. 2015.

25 M. Cha, H. Kwak, P. Rodriguez, Y.-Y. Ahn, and S. Moon. I tube, you tube, everybody tubes: Analyzing the world's largest user generated content video system. In *Proc. ACM SIGCOMM Inter. Measurement Conf.*, pages 1–14, New York, USA, Oct. 2007.

26 F. Benevenuto, T. Rodrigues, M. Cha, and V. Almeida. Characterizing user behavior in online social networks. In *Proc. ACM SIGCOMM Inter. Measurement Conf.*, pages 49–62, New York, USA, Nov. 2009.

27 B. Bai, L. Wang, Z. Han, W. Chen, and T. Svensson. Caching based socially-aware D2D communications in wireless content delivery networks: A hypergraph framework. *IEEE Wirel. Commun.*, 23(4):74–81. 2016.

28 N.-S. Vo, T. Q. Duong, T. D. Hoang, and A. Kortun. Optimal video streaming in dense 5G networks with D2D communications. *IEEE Access*, 6:209–223. 2017.

29 A. Bhardwaj and S. Agnihotri. Energy- and spectral-efficiency trade-off for D2D-multicasts in underlay cellular networks. *IEEE Wirel. Commun. Lett.*, 7(4):546–549. 2018.

30 M.-P. Bui, N.-S. Vo, S. Q. Nguyen, and Q.-N. Tran. Social-aware caching and resource sharing maximized video delivery capacity in 5G ultra-dense networks. *Springer Mobile Netw. Appl.*, 25:2037–2049. 2019.

31 N.-S. Vo, T. Q. Duong, M. Guizani, and A. Kortun. 5G optimized caching and downlink resource sharing for smart cities. *IEEE Access*, 6:31457–31468. 2018.

32 W. Xiang, C. Zhu, C. K. Siew, Y. Xu, and M. Liu. Forward error correction-based 2-D layered multiple description coding for error-resilient H.264 SVC video transmission. *IEEE Trans. Circuits Syst. Video Technol.*, 19(12):1730–1738. 2009.

33 T. Fang and L. P. Chau. GOP-based channel rate allocation using genetic algorithm for scalable video streaming over error-prone networks. *IEEE Trans. Image Process.*, 15(6):1323–1330. 2006.

34 A. Chipperfield, P. Fleming, H. Pohlheim, and C. Fonseca. *Genetic Algorithm TOOLBOX for Use with Matlab Version 1.2 Users Guide*. University of Sheffield, 1994.

35 N.-S. Vo, T. Q. Duong, and M. Guizani. QoE-oriented resource efficiency for 5G two-tier cellular networks: A femtocaching framework. In *Proc. IEEE Global Commun. Conf.*, pages 1–6, Washington, DC, Dec. 2016.

5

Resource Sharing and Multi-tier Multi-rate Caching for Cooperative Video Streaming in Ultra-dense Networks

5.1 Introduction

In 6G UDN,[1] delivering content has become more complex when most of the backhaul links of MBSs and small-cell base stations (SBS) are heavily congested due to a proliferation of MUs requesting an extremely high data transmission rate [2–5]. Caching is often assisted by emerging techniques (millimeter wave, edge computing, multicasting, D2D communications, clustering) [6–10], solutions (resource allocation, cooperative transmission) [10–13], and technologies (massive multiple-input multiple-output, massive ultra-reliable and low latency communications, digital twins) [8, 14] to resolve the congestion situation in the backhaul links.

It is much more challenging to deliver video applications and services, even with three-tier caching deployments [10, 11, 14], because they cannot be applied efficiently. The main reason is that the closely mutual relationship between the complicated features of videos and the characteristics of the wireless environment has not been utilized in designing the system model. Lately, video caching has been extensively studied for 6G networks [15], especially at the edge, e.g., the edge of wireless cellular networks [16] or multi-access edge computing [9]. Video caching is assisted with non-orthogonal transmission, clustering, and multi-access edge computing to maximize the data rate and reduce the transmission delay. However, multi-tier caching has not yet fully enabled flexible video delivery to the MUs from the MBS, SBS, and mobile devices for maximizing playback quality [9, 15, 16].

In this chapter, we propose a resource sharing and multi-tier multi-rate caching (RMC) solution that enables the MUs to receive the video versions flexibly from the MBS (macro tier), the FBSs (femto tier), and the mobile devices (MD tier) in 6G UDN. Each video is encoded into multiple rates with different quality versions. In the macro tier, the MBS is capable of caching all video versions. Meanwhile,

1 This chapter has been published partly in [1].

the video versions are optimally selected to cache in the FBSs depending on the characteristics of wireless channels and the caching storage resource. In the MD tier, the video versions are cached by following a popularity-based probabilistic manner. The MUs are categorized into three types, including spectrum helping MUs (SHs) to share their downlink spectrum resources for D2D communications, the video helping MUs (VHs) to cache the video versions, and the requesting MUs (RUs). Both SHs and VHs receive the video versions from the MBS and the FBSs, while the RUs further receive the video versions from the VHs over D2D communications.

We formulate an RMC optimization problem for cooperative video streaming from the macro tier, femto tier, and MD tier. The MRC optimization problem is solved for the optimal results of (i) the tripartite sets of SHs, VHs, and RUs for establishing video sessions over D2D video communications from the VHs to the RUs by reusing the downlink resources shared by the SHs, and (ii) the caching placement for selecting the proper video versions to be cached in the right FBSs. To make the RMC method more efficient, we consider the constraints on caching storage of the FBSs, the throughput required by the MUs, and target SINR to guarantee the quality of service (QoS) of the SHs. The aim is to provide the MUs with maximum playback quality of video applications and services while saving the caching storage, conserving system throughput, and limiting the interference effect caused by the transmission of the VHs on the SHs. Simulation results are provided to demonstrate the benefits of the proposed RMC method compared with other benchmarks.

The rest of this chapter is organized as follows. In Section 5.2, we introduce the RMC model for cooperative video streaming. Section 5.3 analyzes the RMC model in detail. Based on this analysis, we formulate and solve the RMC optimization problem in Section 5.4. The performance of the RMC method is evaluated in Section 5.5. Section 5.6 concludes the chapter.

5.2 RMC Model for Cooperative Video Streaming

We consider a RMC model for video applications and services in 6G UDNs as illustrated in Figure 5.1. The model has one MBS in the macro tier, J FBSs in the femto tier, and a number of MUs, including K SHs, N VHs, and M RUs, in the MD tier. There is a set of I popular videos in which video i, $i = 1, 2, \ldots, I$, has popularity r_i (2.1) following Zipf-like distribution [17]. Video i is encoded into V_i different quality versions at multiple encoding rates. The video versions are sent from the MBS to the MUs through conventional cellular transmissions, from the FBSs to the MUs using the channel splitting and F-ALOHA schemes to avoid interferences [18], and from the VHs to the RUs over D2D communications reusing the downlink resources shared by the SHs.

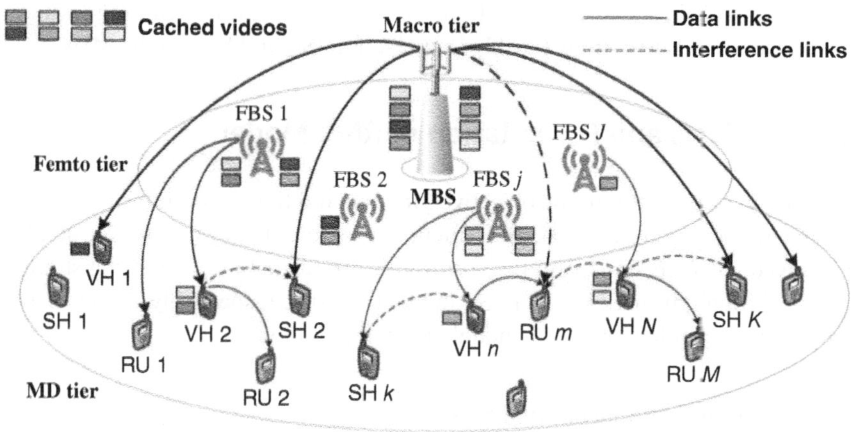

Figure 5.1 An RMC model for video applications and services in 6G UDN.

The spatial distribution of VHs that share the downlink resource from SH k, $k = 1, 2, ..., K$, is modeled as a homogeneous Poisson point process with density $\lambda_C^{(k)}$ [19]. We assume that the system parameters remain at least during the longest streaming session of the considered video set. Under this assumption, the proposed model efficiently serves the MUs the local video applications and services in many crowded areas such as concerts, meeting halls, museums, office buildings, stadiums, hospitals, campuses, etc. Whenever the MBS anticipates that there will be an increasing number of video requests, it implements the RMC method in three steps as follows:

- Step 1—System parameters update: The MBS updates the system parameters, i.e., characteristics of wireless channels; caching storage of FBSs and VHs; features of videos (version, size, and popularity); required throughput of MUs for video playback; and target SINR of SHs.
- Step 2—RMC optimization problem formulation: The MBS formulates the constrained RMC optimization problem and solves it for the optimal results of sharing index $w_{n,m}^{(k)}$ and caching index $u_j^{(v_l)}$. $w_{n,m}^{(k)}$ indicates whether SH k shares its downlink resource with VH n to send video version v_l to RU m $\left(w_{n,m}^{(k)} = 1\right)$ or not $\left(w_{n,m}^{(k)} = 0\right)$, $n = 1, 2, ..., N$, $m = 1, 2, ..., M$, and $v_l = 1, 2, ..., V_l$. Video version v_l is cached in VH n with probability $p_n^{(v_l)}$ depending on the remaining storage of the VH and the size and popularity of the video version. $u_j^{(v_l)}$ indicates whether the FBS j caches video version $v_l\left(u_j^{(v_l)} = 1\right)$ or not $\left(u_j^{(v_l)} = 0\right)$, $j = 1, 2, ..., J$.

- Step 3—Video streaming: Finally, the MBS coordinates all streaming sessions of cached video versions flexibly from itself, FBSs, and VHs to the MUs.

5.3 Analysis and Formulation of RMC Model

This section aims to obtain the objective function and the constraints of the RMC optimization problem. The objective function is represented by the average quality of received video versions, and the constraints are related to the caching storage and the throughput used in the FBSs and by the MUs, respectively. To do so, we analyze and formulate the RMC model in detail as follows.

5.3.1 Preliminary Analysis and Formulation

5.3.1.1 Achievable Rate Probability at SHs

SH k can receive video version v_i from the FBS j or the MBS. If the video version v_i is cached in FBS j, the capacity at SH k is given by

$$C_j^{(k,v_i)} = W \log_2\left(1 + \frac{u_j^{(v_i)} P_j G_j^{(k)}}{\sigma_0}\right), \tag{5.1}$$

however, if none of FBSs caches video version v_i, SH k is normally served by the MBS with the capacity of

$$C_0^{(k,v_i)} = W \log_2\left(1 + \frac{\prod_{j=1}^{J}(1 - u_j^{(v_i)}) P_0 G_0^{(k)}}{\sigma_0 + I_{\mathrm{V,S}}^{(k,v_i)}}\right), \tag{5.2}$$

where W is the system bandwidth; P_j and P_0 are the transmission powers of FBS j and MBS; $G_j^{(k)}$ and $G_0^{(k)}$, which are the channel gains from FBS j and MBS to SH k, are modeled as Rayleigh fading, independently and identically exponentially distributed with unit mean, and multiplied by a standard power law path loss function with a path loss exponent η; σ_0 is the power of additive white Gaussian noise; and $I_{\mathrm{V,S}}^{(k,v_i)}$ is the interference generated by the VHs reusing the same downlink resource share by SH k expressed as

$$I_{\mathrm{V,S}}^{(k,v_i)} = \sum_{n=1}^{N} \sum_{m=1}^{M} w_{n,m}^{(k)} p_n^{(v_i)} P_{\mathrm{V}}^{(k)} G_n^{(k)}, \tag{5.3}$$

where $P_{\mathrm{V}}^{(k)}$ is the transmission power of the VHs associated with SH k, $G_n^{(k)}$ is the channel gain between VH n and SH k, and $p_n^{(v_i)}$ is the probability that VH n caches video version v_i [20], computed as

$$p_n^{(v_i)} = ar_i + b\theta_n^{(v_i)}, \tag{5.4}$$

where $a, b \in [0, 1]$ and $a + b = 1$ are used to balance the impact of popularity and size of video version v_i in making a caching decision. While the impact of popularity is represented by r_i, the impact of size is given by

$$\theta_n^{(v_i)} = r_n \left(1 - \frac{L_i^{(v_i)} - \min\{L_i^{(v_i)}, v_i = 1, 2, \ldots, V_i\}}{L_i^{(v_i)}}\right), \tag{5.5}$$

where r_n is the percentage of remaining storage of VH n and $L_i^{(v_i)}$ is the size of video version v_i.

Let $R_{\text{th}}^{(v_i)}$ be the rate threshold (measured in Kbps) required to send video version v_i for playback, the achievable rate probabilities at SH k provided by FBS j and MBS are, respectively, given by Bhardwaj et al. [19]

$$p_j^{(k,v_i)} = \Pr\left\{C_j^{(k,v_i)} \geq R_{\text{th}}^{(v_i)}\right\} = \exp\left(-\xi_j^{(k,v_i)}\sigma_0/u_j^{(v_i)}P_j\right), \tag{5.6}$$

$$p_0^{(k,v_i)} = \Pr\left\{C_0^{(k,v_i)} \geq R_{\text{th}}^{(v_i)}\right\} = \exp\left\{-\xi_0^{(k,v_i)}\left[\lambda_V^{(k,v_i)}\left(\frac{P_V^{(k)}}{\prod_{j=1}^{J}(1 - u_j^{(v_i)})P_0}\right)^{\frac{2}{\eta}}\right]\right\}, \tag{5.7}$$

where $\xi_j^{(k,v_i)} = (d_j^{(k)})^\eta(2^{R_{\text{th}}^{(v_i)}/W} - 1)$, $\xi_0^{(k,v_i)} = \pi(d_0^{(k)})^2\Gamma(1 + \frac{2}{\eta})\Gamma(1 - \frac{2}{\eta})\ (2^{R_{\text{th}}^{(v_i)}/W} - 1)^{2/\eta}$, $d_j^{(k)}$ and $d_0^{(k)}$ are the distances from FBS j and MBS to SH k, and $\lambda_V^{(k,v_i)} = \sum_{n=1}^{N}\sum_{m=1}^{M} w_{n,m}^{(k)} p_n^{(v_i)}$ is the density of VHs within a circular cell area.

So far, the achievable rate probability to send video version v_i from the FBS j and MBS to SH k is mutually expressed as

$$p_{0,j}^{(k,v_i)} = 1 - \left(1 - p_j^{(k,v_i)}\right)\left(1 - p_0^{(k,v_i)}\right). \tag{5.8}$$

5.3.1.2 Achievable Rate Probability at VHs

Similar to the capacity at the SHs but without interference, the capacities at VH n over the channels from FBS j and MBS are, respectively, expressed as

$$C_{j,n}^{(v_i)} = W\log_2\left(1 + u_j^{(v_i)}P_jG_j^{(n)}/\sigma_0\right), \tag{5.9}$$

$$C_{0,n}^{(v_i)} = W\log_2\left(1 + \frac{\prod_{j=1}^{J}(1 - u_j^{(v_i)})P_0G_0^{(n)}}{\sigma_0}\right), \tag{5.10}$$

where $G_j^{(n)}$ and $G_0^{(n)}$ are the channel gains from FBS j and MBS to VH n.

Then, we can obtain the achievable rate probabilities in accordance with the capacities at VH n, as follows:

$$p_{j,n}^{(v_i)} = \Pr\left\{C_{j,n}^{(v_i)} \geq R_{\text{th}}^{(v_i)}\right\} = \exp\left(\frac{-\xi_j^{(n,v_i)}\sigma_0}{u_j^{(v_i)}P_j}\right), \tag{5.11}$$

$$p_{0,n}^{(v_i)} = \Pr\left\{C_{0,n}^{(v_i)} \geq R_{\text{th}}^{(v_i)}\right\} = \exp\left(\frac{-\xi_0^{(n,v_i)}\sigma_0}{\prod_{j=1}^{J}(1-u_j^{(v_i)})P_0}\right), \tag{5.12}$$

where $\xi_j^{(n,v_i)} = (d_j^{(n)})^\eta(2^{\frac{R_{\text{th}}^{(v_i)}}{W}}-1)$ and $\xi_0^{(n,v_i)} = (d_0^{(n)})^\eta(2^{\frac{R_{\text{th}}^{(v_i)}}{W}}-1)$, $d_j^{(n)}$ and $d_0^{(n)}$ are the distances from FBS j and MBS to VH n.

Therefore, the mutual achievable rate probability to send video version v_i from FBS j and MBS to VH n is

$$p_{0,j,n}^{(v_i)} = 1-\left(1-p_{j,n}^{(v_i)}\right)\left(1-p_{0,n}^{(v_i)}\right). \tag{5.13}$$

5.3.1.3 Achievable Rate Probability at RUs

Different from the capacities at SHs and VHs, the capacity at RU m, which comes from the VHs, FBSs, and MBS, is, respectively, computed as

$$C_{n,m}^{(k,v_i)} = W\log_2\left(1+\frac{w_{n,m}^{(k)}p_n^{(v_i)}P_{\text{V}}^{(k)}G_n^{(m)}}{\sigma_0+P_0G_0^{(m)}+I_{n,m}^{(k,v_i)}}\right), \tag{5.14}$$

$$C_{j,m}^{(k,v_i)} = W\log_2\left(1+\frac{u_j^{(v_i)}\left(1-w_{n,m}^{(k)}p_n^{(v_i)}\right)P_jG_j^{(m)}}{\sigma_0}\right), \tag{5.15}$$

$$C_{0,m}^{(k,v_i)} = W\log_2\left(1+\frac{\prod_{j=1}^{J}\left(1-u_j^{(v_i)}\right)\left(1-w_{n,m}^{(k)}p_n^{(v_i)}\right)P_0G_0^{(m)}}{\sigma_0}\right), \tag{5.16}$$

where $G_n^{(m)}$, $G_j^{(m)}$, and $G_0^{(m)}$ are the channel gains from VH n, FBS j, and MBS to RU m, and the interference caused by the other pairs of VHs and RUs to RU m due to reusing the same spectrum resource shared by SH k is given by

$$I_{n,m}^{(k,v_i)} = \sum_{\substack{n'=1\\n'\neq n}}^{N}\sum_{\substack{m'=1\\m'\neq m}}^{M} w_{n',m'}^{(k)}p_{n'}^{(v_i)}P_{\text{V}}^{(k)}G_{n',m}. \tag{5.17}$$

Thus, the corresponding capacity success probabilities are

$$p_{n,m}^{(k,v_i)} = \Pr\left\{C_{n,m}^{(k,v_i)} \geq R_{\text{th}}^{(v_i)}\right\} \tag{5.18}$$

$$= \exp\left\{-\xi_{n,m}^{(v_i)}\left[\lambda_{\text{M}}\left(\frac{P_0}{w_{n,m}^{(k)} p_n^{(v_i)} P_{\text{V}}^{(k)}}\right)^{\frac{2}{\eta}} + \lambda_{\text{V}}^{\prime(k,v_i)}\right]\right\},$$

$$p_{j,m}^{(k,v_i)} = \Pr\left\{C_{j,m}^{(k,v_i)} \geq R_{\text{th}}^{(v_i)}\right\} = \exp\left[\frac{-\xi_{j,m}^{(v_i)}\sigma_0}{u_j^{(v_i)}\left(1 - w_{n,m}^{(k)} p_n^{(v_i)}\right) P_j}\right], \tag{5.19}$$

$$p_{0,m}^{(k,v_i)} = \Pr\left\{C_{0,m}^{(k,v_i)} \geq R_{\text{th}}^{(v_i)}\right\} = \exp\left[\frac{-\xi_{0,m}^{(v_i)}\sigma_0}{\prod_{j=1}^{J}\left(1 - u_j^{(v_i)}\right)\left(1 - w_{n,m}^{(k)} p_n^{(v_i)}\right) P_0}\right], \tag{5.20}$$

where (5.18) is similarly computed in [19], $\xi_{n,m}^{(v_i)} = \pi(d_n^{(m)})^2\Gamma(1 + \frac{2}{\eta})\Gamma(1 - \frac{2}{\eta})(2^{\frac{R_{\text{th}}^{(v_i)}}{W}} - 1)^{2/\eta}$, $\xi_{j,m}^{(v_i)} = (d_j^{(m)})^\eta(2^{\frac{R_{\text{th}}^{(v_i)}}{W}} - 1)$, and $\xi_{0,m}^{(v_i)} = (d_0^{(m)})^\eta(2^{\frac{R_{\text{th}}^{(v_i)}}{W}} - 1)$; $d_n^{(m)}$, $d_j^{(m)}$, and $d_0^{(m)}$ are the distances from VH n, FBS j, and MBS to RU m; and the density $\lambda_{\text{M}} = 1$ (i.e., one MBS) and $\lambda_{\text{V}}^{\prime(k,v_i)} = \sum_{\substack{n'=1 \\ n' \neq n}}^{N} \sum_{\substack{m'=1 \\ m' \neq m}}^{M} w_{n',m'}^{(k)} p_{n'}^{(v_i)}$ within a circular cell area.

Finally, the mutual achievable rate probability at RU m is

$$p_{0,j,n,m}^{(k,v_i)} = 1 - \left(1 - p_{n,m}^{(k,v_i)}\right)\left(1 - p_{j,m}^{(k,v_i)}\right)\left(1 - p_{0,m}^{(k,v_i)}\right). \tag{5.21}$$

5.3.2 Objective Function

Based on the achievable rate probabilities in (5.8), (5.13), and (5.21), we can compute the overall average playback quality of received video versions at the MUs (i.e., SHs, VHs, and RUs), which is considered as the objective function of RMC optimization problem. To do so, we apply the rate–distortion model (RDM) expressed by a decaying exponential function [21], which accurately represents the experimental RD characteristics of the videos encoded by high-efficiency video coding standard [22]. In this model, if video version v_i is encoded and then played back at rate $R_{\text{th}}^{(v_i)}$, the reconstructed distortion is

$$D_i\left(R_{\text{th}}^{(v_i)}\right) = \gamma_i\left(R_{\text{th}}^{(v_i)}\right)^{\beta_i}, \tag{5.22}$$

where γ_i and β_i are the sequence-dependent parameters selected so that (5.22) meets the experimental rate–distortion curves.

The overall average playback quality of received video versions at the SHs, VHs, and RUs is expressed as

$$\overline{Q} = \frac{1}{3J} \sum_{j=1}^{J} \left(Q_{\mathrm{S}}^{(j)} + Q_{\mathrm{V}}^{(j)} + Q_{\mathrm{R}}^{(j)} \right), \tag{5.23}$$

where

$$Q_{\mathrm{S}}^{(j)} = \frac{1}{K} \sum_{k=1}^{K} \sum_{i=1}^{I} \frac{r_i}{V_i} \sum_{v_i=1}^{V_i} p_{0,j}^{(k,v_i)} Q\left(D_i\left(R_{\mathrm{th}}^{(v_i)}\right)\right), \tag{5.24}$$

$$Q_{\mathrm{V}}^{(j)} = \frac{1}{N} \sum_{n=1}^{N} \sum_{i=1}^{I} \frac{r_i}{V_i} \sum_{v_i=1}^{V_i} p_{0,j,n}^{(v_i)} Q\left(D_i\left(R_{\mathrm{th}}^{(v_i)}\right)\right), \tag{5.25}$$

$$Q_{\mathrm{R}}^{(j)} = \frac{1}{KMN} \sum_{k=1}^{K} \sum_{N=1}^{N} \sum_{m=1}^{M} \sum_{i=1}^{I} \frac{r_i}{V_i} \sum_{v_i=1}^{V_i} p_{0,j,n,m}^{(k,v_i)} Q\left(D_i\left(R_{\mathrm{th}}^{(v_i)}\right)\right). \tag{5.26}$$

It is noted that in (5.24), (5.25), and (5.26), $Q\left(D_i\left(R_{\mathrm{th}}^{(v_i)}\right)\right) = 10 \log_{10} \frac{255^2}{D_i\left(R_{\mathrm{th}}^{(v_i)}\right)}$ is the peak signal-to-noise ratio (PSNR) is measured in dB.

5.3.3 Resource Consumption

In the RMC model for cooperative video streaming, the resource consumption includes the total storage of the FBSs ($L_{\mathrm{F}}^{(i)}$) consumed to cache all versions of video i and the throughput requested by the MUs (C_{T}), which are, respectively, given by

$$L_{\mathrm{F}}^{(i)} = \sum_{j=1}^{J} \sum_{v_i=1}^{V_i} u_j^{(v_i)} L_i^{(v_i)}, \tag{5.27}$$

$$C_{\mathrm{T}} = C_{\mathrm{S}} + C_{\mathrm{V}} + C_{\mathrm{R}}, \tag{5.28}$$

where

$$C_{\mathrm{S}} = \sum_{j=1}^{J} \sum_{k=1}^{K} \sum_{i=1}^{I} \frac{r_i}{V_i} \sum_{v_i=1}^{V_i} \left(C_j^{(k,v_i)} + C_0^{(k,v_i)} \right), \tag{5.29}$$

$$C_{\mathrm{V}} = \sum_{j=1}^{J} \sum_{n=1}^{N} \sum_{i=1}^{I} \frac{r_i}{V_i} \sum_{v_i=1}^{V_i} \left(C_{j,n}^{(v_i)} + C_{0,n}^{(v_i)} \right), \tag{5.30}$$

$$C_{\mathrm{R}} = \sum_{j=1}^{J} \sum_{k=1}^{K} \sum_{N=1}^{N} \sum_{m=1}^{M} \sum_{i=1}^{I} \frac{r_i}{V_i} \sum_{v_i=1}^{V_i} \left(C_{n,m}^{(k,v_i)} + C_{j,m}^{(k,v_i)} + C_{0,m}^{(k,v_i)} \right). \tag{5.31}$$

5.4 RMC Optimization Design

So far, aiming at maximizing the objective function (5.23), letting $L_{\max}^{(i)}$, C, and γ_0 be the constraints on the caching storage of FBSs to save, the throughput required by MUs to conserve, and the target SINR to guarantee the QoS of SHs, the RMC optimization problem is formulated as

$$\max_{u_j^{(v_i)}, w_{n,m}^{(k)}} \overline{Q}, \tag{5.32a}$$

$$\text{s.t.} \sum_{v_i=1}^{V_i} u_j^{(v_i)} \leq 1, i = 1, 2, \ldots, I, j = 1, 2, \ldots, J, \tag{5.32b}$$

$$\sum_{m=1}^{M} w_{n,m}^{(k)} \leq 1, k = 1, 2, \ldots, K, n = 1, 2, \ldots, N, \tag{5.32c}$$

$$\sum_{n=1}^{N} w_{n,m}^{(k)} \leq 1, k = 1, 2, \ldots, K, m = 1, 2, \ldots, M, \tag{5.32d}$$

$$L_{\mathrm{F}}^{(i)} \leq \mu L_{\max}^{(i)}, 0 < \mu \leq 1, i = 1, 2, \ldots, I \tag{5.32e}$$

$$C_{\mathrm{T}} \leq \delta C, 0 < \delta \leq 1, \tag{5.32f}$$

$$I_{\mathrm{V,S}}^{(k,v_i)} \leq \frac{P_0 G_0^{(k)}}{\gamma_0} - \sigma_0, k = 1, 2, \ldots, K, i = 1, 2, \ldots, I, v_i = 1, 2, \ldots, V_i, \tag{5.32g}$$

where the constraint (5.32b) is to limit the FBS j to cache up to one proper version of video i. The constraints (5.32c) and (5.32d) are to guarantee that VH n is able to transmit to up to one RU and RU m is able to receive from up to one VH. μ (5.32e) and δ (5.32f) are to flexibly adjust $L_{\max}^{(i)}$ and C for convenience of evaluation; here $L_{\max}^{(i)} = r_i \times J \times I \times \max\left\{L_i^{(v_i)}, v_i = 1, 2, \ldots, V_i\right\}$. The last constraint (5.32g), which comes from (5.2), is to limit the effect of interference on the QoS of every SH.

The above RMC optimization problem can be solved for the optimal results of $u_j^{(v_i)}$ and $w_{n,m}^{(k)}$ by applying the exhaustive binary search (EBS) algorithm [20]. Although the memory and time complexity of the EBS algorithm are much higher than other methods, e.g., stochastic global search methods like genetic algorithms for exact or approximated optimal results, especially in the large scale of 6G UDN, it is acceptable due to the following two reasons. First, it is a simple method to find the optimal global results. Second, the proposed RMC method is efficiently used to provide the MUs with local video applications and services in crowded areas, in which the system characteristics change mainly depending on the arrival of MUs following a daily pattern with stability on an hourly basis [23]. This pattern gives us the opportunity to relax the strict requirement of time complexity; meanwhile, the memory resource can be powerfully assisted by the MBS. In addition, to ensure that SH k can share its downlink resource with only one pair of VH n and RU m, the EBS algorithm follows the rule that if $w_{n,m}^{(k)} = 1$, then $w_{n,m}^{(k')} = 0, k' = 1, 2, \ldots, K, k' \neq k$.

5.5 Performance Evaluation of RMC

The RMC model is deployed with the parameters listed in Table 5.1. Assume that the MBS covers the circular cell area within a radius of 1,500 m. We consider 3 videos, i.e., $V_1^{(v_1)}$ (Basketballpass), $V_2^{(v_2)}$ (Racehorses), and $V_3^{(v_3)}$ (Foreman), to analyze their experimental RD curves by using HM reference software version 12.0 [24] and then obtain $L_i^{(v_i)}$, $R_{\text{th}}^{(v_i)}$, γ_i, and β_i. To evaluate the performance of RMC, we compare it to the other three schemes, including only caching (OCC), only resource sharing (ORS), and no caching nor resource sharing (NCS).

Table 5.1 Parameters setting.

Symbols	Values
J, K, N, M	3 FBSs, 3 SHs, 4 VHs, 5 RUs
I, V_i	3 videos, each has 3 versions
$P_0, P_j, P_C^{(k)}, \sigma_0$	5 W, 1 W, 10^{-3} W, 10^{-12}W
$W, \gamma_0, \eta, \alpha$	5 MHz, 3 dB, 4, 1
a, b, μ, δ	0.5, 0.5, 0.5, 1
r_n	Fixed to 1, i.e., all CUs have 100% of storage to cache
$L_i^{(v_i)}$	[11,867 23,734 35,600; 198,680 264,906 351,000; 33,382 66,763 113,496] Kbit
$R_{\text{th}}^{(v_i)}$	[1,000 2,000 3,000; 3,000 4,000 5,300; 50,000 100,000 170,000] Kbps
$\{\gamma_i; \beta_i\}$	{9,806 76,520 1,644,000;−0.9972 −1.1530 −1.0920}
C	12 Gbps, i.e., each MU is served at 1 Gbps
Distances	**Specifications**
MBS to MUs	Randomly uniform distributed in [500, 1,500] m
FBSs to MUs	Randomly uniform distributed in [21, 100] m
VHs to SHs	Randomly uniform distributed in [50, 150] m
VHs to RUs	Randomly uniform distributed in [2, 21] m

Figure 5.2 shows the performance of RMC, OCC, ORS, and NCS versus the skewed popularity exponent α. The increase of α makes the popularity more skewed amongst the videos, i.e., yielding a smaller number of high popular videos but a larger number of low-popular ones. In this context, the system focuses on serving more MUs the videos with higher popularity rather than the ones with lower popularity to gain higher performance. The proposed RMC outperforms the other OCC, ORS, and NCS thanks to the joint method of caching technique assisted by resource sharing and adaptive rate encoding selection. The OCC is mostly better than the ORS because it is likely to provide more possibilities of caching and transmitting over better channels from FBSs to MUs than the ORS over the D2D communications. The NCS is the worst scheme without RMC assistance.

In Figure 5.3, we investigate the effect of the constraints on the performance of RMC by considering the other three cases of constraints, including (i) RMC_1: decreasing the caching storage of FBSs by changing μ in (5.32e) from 0.5 to 0.3, (ii) RMC_2: decreasing the required throughput of MUs by changing δ in (5.32f)

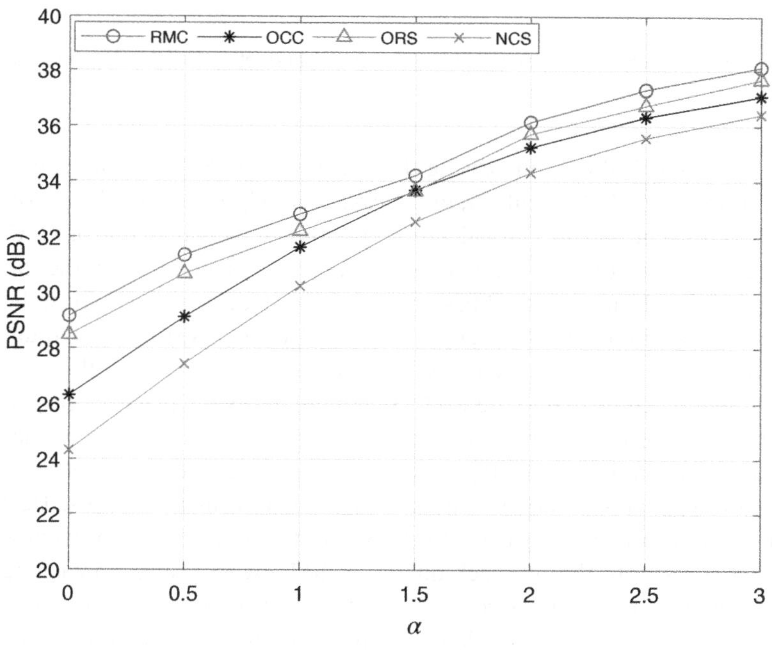

Figure 5.2 Playback quality versus α. *Source*: Reproduced from [1] / IEEE.

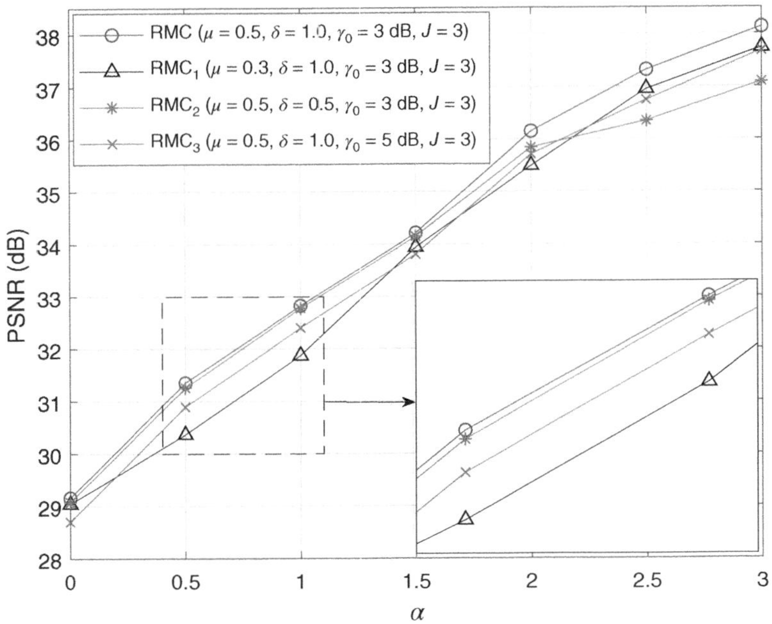

Figure 5.3 Performance of RMC versus different system parameters. *Source*: Reproduced from [1] / IEEE.

from 1 to 0.5, and (iii) RMC$_3$: increasing the target SINR of SHs by changing γ_0 in (5.32g) from 3 to 5 dB. Particularly, in the case of RMC$_1$, the system has fewer opportunities to cache and thus provides lower playback quality than that of RMC. In RMC$_2$, the required throughput of MUs decreases; for example, when the statistical estimates indicate that the MUs are with lower playback resolutions, the wireless channels with higher throughput are not selected for streaming in order to conserve system bandwidth resources. Therefore, the system serves the MUs with lower quality (compared to the RMC) that quite meets the playback resolutions of the MUs. In RMC$_3$, when the target SINR γ_0 increases to ensure higher QoS for the SHs, the number of D2D pairs is reduced to mitigate the effect of interference on the SHs. This situation reduces the system performance because D2D communications cannot be fully exploited for video streaming in close proximity.

To further investigate the playback quality of RMC with respect to caching storage consumed in the FBSs, we compare it to the full rate caching and resource sharing scheme (FRS). In FRS, the deployment follows the same manner as the RMC, but the FBSs always cache the video versions with the highest encoding rates, i.e., without adaptive rate encoding selection. The results in Figure 5.4 show that the FRS outperforms the RMC when $\alpha > 1$. The reason is that when α gets

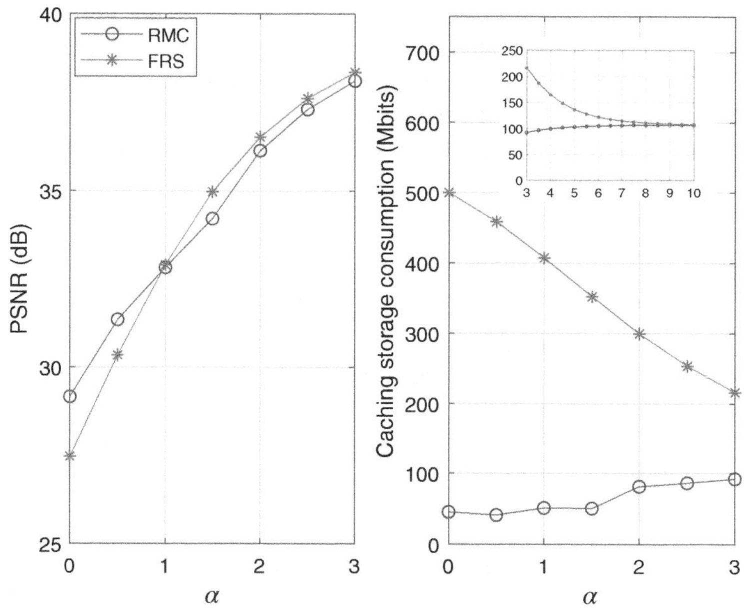

Figure 5.4 Performance of RMC and FRS. *Source*: Reproduced from [1] / IEEE.

higher values, the system focuses on the videos with higher popularity to serve the MUs. In this context, while the performance of RMC is limited by the caching storage and required throughput constraints, these constraints are fully relaxed in the FRS. In consideration of caching storage consumption, the FRS clearly uses higher storage than the RMC does. The caching storage consumption of both RMC and FRS converges on a certain value with respect to the increase of α because the smaller number of high popular videos with the highest encoding rates selected to cache is the same in both RMC and FRS. To make clear about the convergence, a further result of caching storage consumption of RMC and FRS versus α extended from 3 to 10 is plotted in the top right corner of Figure 5.4.

5.6 Conclusion

We have proposed the RMC method that applies not only the caching and underlay D2D communication techniques but also the resource sharing and adaptive rate encoding selection solutions to provide the MUs with video applications and services in 6G UDN. By utilizing the caching storage resource of MBS, FBS, and VH and the downlink spectrum resource of SH, the RMC optimization problem was formulated for finding the optimal results to cache the video versions in the FBS and share the downlink resource of SH. The constraints on caching storage of the FBS, required throughput of the MU, and target SINR of the SH were considered to improve the efficiency of the RMC method. Simulation results demonstrated the benefits of the RMC compared to other schemes in terms of playback quality and caching storage consumption.

References

1 N.-S. Vo, M.-P.Bui, P. Q. Truong, C. Yin, and A. Masaracchia. Multi-tier caching and resource sharing for video streaming in 5G ultra-dense networks. *IEEE Commun. Lett.*, 24(7):1500–1504. 2020.

2 T. H. Nguyen, D. Q. Nguyen, and V. D. Nguyen. Quality of service provisioning for D2D users in heterogeneous networks. *EAI Endorsed Trans. Ind. Netw. Intell. Syst.*, 6(21):1–7. 2019.

3 M. T. Nguyen. An energy-efficient framework for multimedia data routing in Internet of things (IoTs). *EAI Endorsed Trans. Ind. Netw. Intell. Syst.*, 6(19):1–8. 2019.

4 H. T. Nguyen, H. D. Tuan, T. Q. Duong, H. V. Poor, and W.-J. Hwang. Collaborative multicast beamforming for content delivery by cache-enabled ultra dense networks. *IEEE Trans. Commun.*, 67(5):3396–3406. 2019.

5 N.-S. Vo, T. Q. Duong, and M. Guizani. QoE-oriented resource efficiency for 5G two-tier cellular networks: A femtocaching framework. In *Proc. IEEE Global Commun. Conf.*, pages 1–6, Washington, DC, Dec. 2016.

6 Z. Lin, Y. Fang, P. Chen, F. Chen, and G. Zhang. Modeling and analysis of edge caching for 6G mm Wave vehicular networks. *IEEE Trans. Intell. Transport. Syst.*, 24(7):7422–7434. 2023.

7 T. M. Ayenew, D. Xenakis, L. Alonso, N. Passas, and L. Merakos. Demand-aware cooperative content caching in 5G/6G networks with MEC-enabled edges. *IEEE Netw. Lett.*, 4(3):118–122. 2022.

8 M. Zhou, J. Li, J. Yuan, M. Xie, W. Tan, and R. Yin. An architecture for AoI and cache hybrid multicast/unicast/D2D with cell-free massive MIMO systems. *IEEE Access*, 11:43080–43088. 2023.

9 A. Doostmohammadi, M. R. Khayyambashi, N. Movahedinia, and Z. Becvar. Dynamic clustering for low-delay delivery of video content cached in MEC servers. *IEEE Syst. J.*, 17(4):5842–5853. 2023.

10 Q. Liu, H. Zhang, X. Zhang, and D. Yuan. Joint service caching, communication and computing resource allocation in collaborative MEC systems: A DRL-based two-timescale approach. *IEEE Trans. Wirel. Commun.*, 23(10):15493–15506. 2024.

11 X. Li, X. Wang, K. Li, Z. Han, and V. C. M. Leung. Collaborative multi-tier caching in heterogeneous networks: Modeling, analysis, and design. *IEEE Trans. Wirel. Commun.*, 16(10):6926–6939. 2017.

12 D.-H. Tran, S. Chatzinotas, and B. Ottersten. Satellite- and cache-assisted UAV: A joint cache placement, resource allocation, and trajectory optimization for 6G aerial networks. *IEEE Open J. Veh. Technol.*, 3:40–54. 2022.

13 T. Feng, S. Shi, S. Gu, N. Zhang, W. Xiang, and X. Gu. Performance analysis for cache-enabled cellular networks with cooperative transmission. In *Proc. IEEE Veh. Technol. Conf.*, pages 1–5, Victoria, BC, Canada, Nov. 2020.

14 X. Zhang, Q. Zhu, and H. V. Poor. Multi-tier caching for statistical-QoS driven digital twins over mURLLC-based 6G massive-MIMO mobile wireless networks using FBC. *IEEE J. Sel. Topics Signal Process.*, 18(1):34–49. 2024.

15 J. Ma, L. Liu, B. Shang, S. Jere, and P. Fan. Performance analysis and optimization for layer-based scalable video caching in 6G networks. *IEEE/ACM Trans. Netw.*, 31(4):1494–1506. 2023.

16 X. Zhang, Y. Ren, T. Lv, and L. Hanzo. Caching scalable videos in the edge of wireless cellular networks. *IEEE Netw.*, 37(3):34–42. 2023.

17 L. Breslau, P. Cao, L. Fan, G. Phillips, and S. Shenker. Web caching and Zipf-like distributions: Evidence and implications. In *Proc. IEEE Inter. Conf. on Computer Commun.*, pages 126–134, New York, NY, Mar. 1999.

18 W. C. Cheung, T. Q. S. Quek, and M. Kountouris. Throughput optimization, spectrum allocation, and access control in two-tier femtocell networks. *IEEE J. Sel. Areas Commun.*, 30(3):561–574. 2012.

19 A. Bhardwaj and S. Agnihotri. Energy- and spectral-efficiency trade-off for D2D-multicasts in underlay cellular networks. *IEEE Wirel. Commun. Lett.*, 7(4):546–549. 2018.

20 N.-S. Vo, T. Q. Duong, M. Guizani, and A. Kortun. 5G optimized caching and downlink resource sharing for smart cities. *IEEE Access*, 6:31457–31468. 2018.

21 W. Xiang, C. Zhu, C. K. Siew, Y. Xu, and M. Liu. Forward error correction-based 2-D layered multiple description coding for error-resilient H.264 SVC video transmission. *IEEE Trans. Circuits Syst. Video Technol.*, 19(12):1730–1738. 2009.

22 G. J. Sullivan, J.-R. Ohm, W.-J. Han, and T. Wiegand. Overview of the high efficiency video coding (HEVC) standard. *IEEE Trans. Circuits Syst. Video Technol.*, 22(12):1649–1668. 2012.

23 H. Yu, D. Zheng, B. Y. Zhao, and W. Zheng. Understanding user behavior in large-scale video-on-demand systems. In *Proc. ACM EuroSyst.*, page 333–344, Leuven, Belgium, Apr. 2006.

24 HEVC. *HM Reference Software Version 12.0.* http://hevc.hhi.fraunhofer.de. 2013.

6

Social-aware Resource Sharing and Caching for Cooperative Video Transmission in Ultra-dense Networks

6.1 Introduction

In the past decade, we have witnessed the convergence between social networks and VAS.[1] Social behaviors and characteristics of MUs certainly shape the way VAS are distributed through networks [2, 3]. In 6G UDNs, a huge number of MUs with diverse social behaviors and characteristics simultaneously demand a high quality of experience (QoE). On the other hand, VAS consumes a substantial amount of system resources, including storage, spectrum, bandwidth, and energy. Therefore, emerging techniques and solutions for VAS in 6G UDN cannot work efficiently without identifying the social behaviors and characteristics of MUs, the aspects of QoE, and the resource consumption.

Social attributes and social-aware schemes have been utilized to explore caching techniques for VAS [4–8]. In [4], an edge caching scheme for VAS, which can cache layered videos for different social groups of MUs in close connections, was studied, focusing on how to find the optimal number of video layers and the optimal price to cache them for lower transmission delay. In addition to geographical proximity, the work [5] considered the history of encounters to identify the social relationship between the MUs and thus enhance the cache hit and sum rate. A socially aware D2D video delivery method based on mobility similarity measurements was studied in [6] to build a social state transition model of MUs' movement. The proposed social-aware scheme can improve the system performance in terms of packet loss rate, average freeze time, cache utilization, average bitrate, buffer level, and control overhead. However, only D2D communications were considered, while cooperative video transmission was not examined.

1 This chapter has been published partly in [1].

A collaborative social-aware and QoE-driven video caching and adaptation in edge network was studied in [7]. Real-world data was used to verify and model the social network as a dynamic bi-direct graph that can predict the MUs' time-varying social relationships and reduce the gap between theory and reality. A two-tier collaborative multi-access edge computing cache architecture is proposed to maximize the hit ratio and QoE. In [8], socially aware edge caching was applied to unmanned aerial vehicle-assisted metaverse systems. However, cooperative video transmission model has not been studied.

In this chapter, we propose a SRC for cooperative video transmission in 6G UDN. Similar to Chapter 5 in terms of multi-tier caching and cooperative video transmission, but further considering the social relationship between the MUs, we formulate the SRC optimization problem and solve it for the optimal results of (i) a number of caching copies of each video and caching placements in the FBS and (ii) resource sharing between the spectrum helping MUs (SHs) and the D2D pairs for D2D communications. Here, each D2D pair consists of a video helping MU (VH) and a requesting MU (RU). The objective is to maximize the cache-hit ratio and video delivery capacity while satisfying the target SINR of the SHs. Simulation results demonstrate the benefits of the proposed SRC method compared to other conventional schemes that do not consider social relationships nor downlink resource sharing. Our findings also reveal that there exists an appropriate region where the MUs achieve the highest video delivery capacity.

The rest of this chapter is organized as follows. Section 6.2 introduces a system model for cooperative video transmission. The analysis of SRC is presented in Section 6.3. The SRC optimization problem and solution are provided in Section 6.4. Section 6.5 presents simulation results and performance evaluation. Finally, Section 6.6 concludes the chapter.

6.2 SRC Model for Cooperative Video Transmission

We consider an SRC model for cooperative video transmission as illustrated in Figure 6.1. In the device layer, there are one MBS, I videos, J FBSs, K SHs, and N D2D pairs, each with a VH and an RU. In the social layer, each D2D pair has its own social relationship represented by their history contact duration and encounters. The detailed SRC method implemented at the MBS is described in the following three steps:

- Step 1—Update the system parameters: If there are some significant changes, the MBS updates the information of social relationship, device layer (storage of FBSs and VHs and characteristics of wireless channels), and videos (popularity and length).

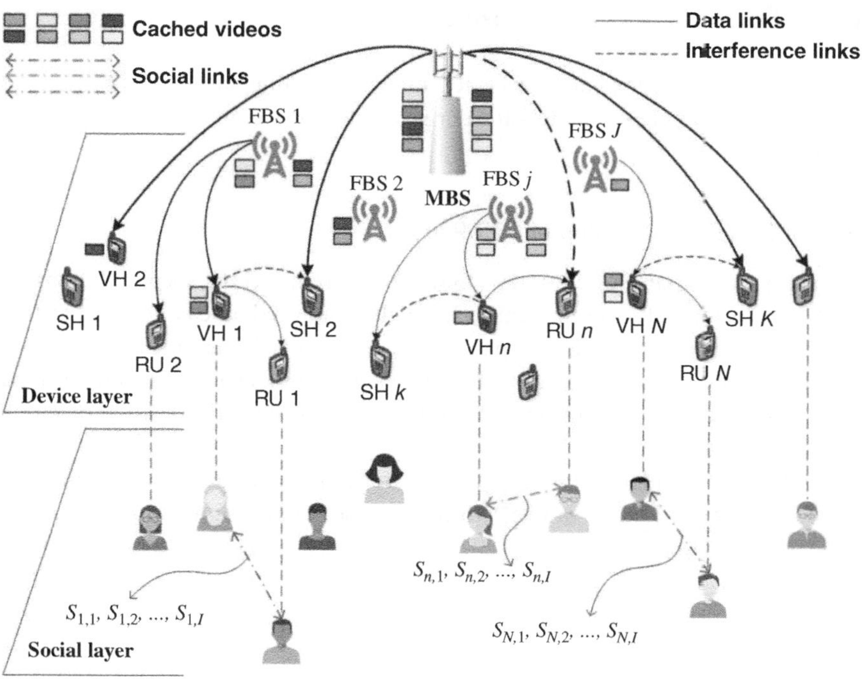

Figure 6.1 SRC model for cooperative video transmission in UDN.

- Step 2—Formulate and solve the SRC optimization problem: The system parameters enable the MBS to formulate the SRC optimization problem and then solve it for the optimal results of (i) a number of caching copies (c_i) of video i and caching placements in the FBSs represented by the caching index $u_{j,i}$, $j = 1, 2, ..., J$ and $i = 1, 2, ..., I$, i.e., to identify that if FBS j caches video i ($u_{j,i} = 1$) or does not ($u_{j,i} = 0$) and (ii) downlink resource sharing between the SHs and the D2D pairs represented by the resource sharing index $v_{k,n}$, $k = 1, 2, ..., K$ and $n = 1, 2, ..., N$, to identify that if SU k shares its downlink resource with D2D pair n ($v_{k,n} = 1$) or does not ($v_{k,n} = 0$). An arbitrary SH can share its downlink resource with more than one D2D pair as long as the target SINR of the SH is guaranteed.
- Step 3—Implement the SRC method: Based on the optimal results, the MBS assigns the FBSs to cache the right videos and the SHs to share their downlink resources with the right D2D pairs, for cooperatively transmitting the videos from the MBS, FBSs, and VHs to the MUs at high system performance of cache-hit ratio and delivery capacity.

6.3 Formulation of SRC Model

In this section, we formulate the SRC model including social relationships of D2D pairs and wireless channels and derive the overall average delivery capacity of the system.

6.3.1 Social Relationship of D2D Pairs

We apply the Indian Buffet Model (IBM) [9–14] to model the social relationship of D2D pairs for D2D communications, which is captured based on their duration and encounter histories. A D2D pair with a closer social relationship provides a higher probability to successfully offload a video. The social relationship of D2D pair n is utilized for deriving the probability to successfully offload video i of length $T^{(i)}_{\min}$ presented in the sequel.

We define Z_m and Z_n, $m = 1, 2, \ldots, Z_n$ as the contact duration and the number of encounters of D2D pair n, respectively. The mean of encounter duration M_n of D2D pair n is given by

$$M_n = \frac{\sum_{m=1}^{Z_n} Z_m}{Z_n}. \tag{6.1}$$

The variance of encounter duration of D2D pair n is expressed as

$$V_n = \frac{\sum_{m=1}^{Z_n} (Z_m - M_n)^2}{Z_n}. \tag{6.2}$$

From (6.1) and (6.2), the encounter duration is modeled as a Gamma distribution as

$$X \sim \Gamma(\kappa_n, \theta_n) = \Gamma(M_n^2/V_n, V_n/M_n) \tag{6.3}$$

and the probability density function of encounter duration is defined as

$$f(x; \kappa_n, \theta_n) = \frac{1}{\theta_n^{\kappa_n}} \frac{1}{\Gamma(\kappa_n)} x^{\kappa_n - 1} e^{-\frac{x}{\theta_n}}, \tag{6.4}$$

where $\Gamma(\kappa_n) = \int_0^\infty t^{\kappa_n - 1} e^{-t} dt$.

The probability that D2D pair n successfully offloads video i of length $T^{(i)}_{\min}$ is computed by

$$s_{n,i} = 1 - \int_0^{\delta T^{(i)}_{\min}} f(u; \kappa_n, \theta_n) du = 1 - \frac{\gamma(\kappa_n, \frac{\delta T^{(i)}_{\min}}{\theta_n})}{\Gamma(\kappa_n)}, \tag{6.5}$$

where $\gamma(\kappa_n, \frac{\delta T_{\min}^{(l)}}{\theta_n}) = \int_0^{\frac{\delta T_{\min}^{(l)}}{\theta_n}} t^{\kappa_n - 1} e^{-t} dt$, and $\delta \geq 1$ is used to adjust the length of all videos to meet the social relationships of all D2D pairs and thus gain the highest system performance.

6.3.2 Wireless Channels

In our system, due to the overlaid presence of both MSB and FSBs, which may cause cross-tier and co-tier interferences, we apply the channel splitting and F-ALOHA schemes to control the interferences [15, 16]. So, there exists only the interference generated from the MBS and VHs to the SHs and RUs reusing the same downlink resources shared by the SHs for D2D communications. The wireless channels are featured by the channel gains [16], expressed as

$$G_{\mathsf{X,Y}}^{\mathsf{x,y}} = h_{\mathsf{X,Y}}^{\mathsf{x,y}} g_{\mathsf{X,Y}}^{\mathsf{x,y}}, \tag{6.6}$$

where $\mathsf{X} \in \{\mathsf{M, F, V}\}$ is the transmitters {MBS, FBS, VH}, $\mathsf{Y} \in \{\mathsf{S, V, R}\}$ is the receivers {SH, VH, RU}, $\mathsf{x} \in \{j, n\}$, $j = 0$ implies the MBS, and $\mathsf{y} \in \{k, n\}$; $h_{\mathsf{X,Y}}^{\mathsf{x,y}}$ is the exponential power fading coefficient; and $g_{\mathsf{X,Y}}^{\mathsf{x,y}} = ||d_{\mathsf{X,Y}}^{\mathsf{x,y}}||^{-\eta}$ is the standard power law path loss function in which η is the path loss exponent, d is the distance between X and Y, and $||.||$ is the Euclidean norm.

6.3.3 System Delivery Capacity

The system delivery capacity is defined as the throughput cooperatively delivered from the MBS, FBSs, and VHs to the SHs, VHs, and RUs. Given the aforementioned channel gains, the corresponding system delivery capacities are presented in the sequel.

6.3.3.1 Capacity Delivered to SHs

The SHs can share their downlink resources while receiving the videos from the MBS and FBSs. When receiving video i cached in FSB j with $u_{j,i} = 1$, the signal-to-noise ratio (SNR) at SH k over the channel from FBS j is given by

$$\gamma_{\mathsf{F,S}}^{(j,k,i)} = \frac{u_{j,i} P_{\mathsf{F}}^{(j)} G_{\mathsf{F,S}}^{(j,k)}}{\sigma_0}. \tag{6.7}$$

where $P_{\mathsf{F}}^{(j)}$ is the transmission power of FBS j, $G_{\mathsf{F,S}}^{(j,k)}$ is the channel gain between FBS j and SH k, and σ_0 is the power of additive white Gaussian noise (AWGN).

Otherwise, if none of the FBSs caches video i, i.e., $u_{j,i} = 0, \forall j$, SH k normally receives video i from the MBS. In this case, it suffers from the interference caused by the transmission of VH n, which reuses the same downlink resource.

Furthermore, this interference effect depends on (i) the probability $p_{n,i}$ that video i is cached in VH n and (ii) the probability $s_{n,i}$ that video i is successfully offloaded to RU n based on the social relationship $s_{n,i}$ of D2D pair n. The SINR at SH k over the channel from the MBS is given by

$$\gamma_{\mathsf{M,S}}^{(0,k,i)} = \frac{\prod_{j=1}^{J}(1-u_{j,i})P_{\mathsf{M}}^{(0)}G_{\mathsf{M,S}}^{(0,k)}}{\sigma_0 + \sum_{n=1}^{N} s_{n,i}v_{k,n}p_{n,i}P_{\mathsf{V}}^{(n)}G_{\mathsf{V,S}}^{(n,k)}}. \tag{6.8}$$

In (6.8), $P_{\mathsf{M}}^{(0)}$ and $P_{\mathsf{V}}^{(n)}$ are the transmission powers of the MBS and VH n, $G_{\mathsf{M,S}}^{(0,k)}$ is the channel gain between the MBS and SH k, and similar to (5.4), the probability $p_{n,i}$ of VH n to cache video i relies on the popularity of video i (r_i) and the percentage of available storage of VH n (β_n) defined as

$$p_{n,i} = ar_i + b\beta_n. \tag{6.9}$$

The average capacity of all videos delivered to the SHs can be obtained as

$$R_{\mathsf{S}} = W\left\{\sum_{i=1}^{I} r_i \sum_{k=1}^{K}\left[\sum_{j=1}^{J}\log_2(1+\gamma_{\mathsf{F,S}}^{(j,k,i)}) + \log_2(1+\gamma_{\mathsf{M,S}}^{(0,k,i)})\right]\right\}, \tag{6.10}$$

where W is the system bandwidth.

6.3.3.2 Capacity Delivered to VHs

The VHs can share their caching storage while receiving the videos from the MBS and FBSs. When receiving video i cached in FSB j with $u_{j,i} = 1$, the SNR at VH n over the channel from FBS j is given by

$$\gamma_{\mathsf{F,V}}^{(j,n,i)} = \frac{u_{j,i}P_{\mathsf{F}}^{(j)}G_{\mathsf{F,V}}^{(j,n)}}{\sigma_0}. \tag{6.11}$$

Otherwise, if none of the FBSs cache video i, i.e., $u_{j,i} = 0, \forall j$, VH n is served by the MBS with the SINR simply given by

$$\gamma_{\mathsf{M,V}}^{(0,n,i)} = \frac{\prod_{j=1}^{J}(1-u_{j,i})P_{\mathsf{M}}^{(0)}G_{\mathsf{M,V}}^{(0,n)}}{\sigma_0}, \tag{6.12}$$

where $G_{\mathsf{F,V}}^{(j,n)}$ and $G_{\mathsf{M,V}}^{(0,n)}$ are the channel gains from FBS j and MBS to VH n.

Based on (6.11) and (6.12), the average capacity of all videos delivered to the VHs can be obtained as

$$R_{\mathsf{T}} = W\left\{\sum_{i=1}^{I} r_i \sum_{n=1}^{N}\left[\sum_{j=1}^{J}\log_2(1+\gamma_{\mathsf{F,V}}^{(j,n,i)}) + \log_2(1+\gamma_{\mathsf{M,V}}^{(0,n,i)})\right]\right\}. \tag{6.13}$$

6.3.3.3 Capacity Delivered to RUs

The capacity delivered to the RUs comes from not only the MBS and FBSs, but also the VHs over D2D communications reusing the downlink resources shared by the SHs. By further considering the three conditions for (i) the social relationship based successful probability to offload video i over D2D pair n ($s_{n,i}$), (ii) the downlink resource sharing index of SH k with D2D pair n ($v_{k,n}$), and (iii) the caching probability of video i at VH n ($p_{n,i}$), the SINR at RU n over the channel from VH n is given by

$$\gamma_{\mathrm{V,R}}^{(n,k,i)} = \frac{s_{n,i} v_{k,n} p_{n,i} P_{\mathrm{V}}^{(n)} G_{\mathrm{V,R}}^{(n,n)}}{\sigma_0 + P_{\mathrm{M}}^{(0)} G_{\mathrm{M,R}}^{(0,n)} + \sum_{l=1,l\neq n}^{N} s_{l,i} v_{k,l} p_{l,i} P_{\mathrm{V}}^{(l)} G_{\mathrm{V,R}}^{(l,l)}}, \tag{6.14}$$

where $G_{\mathrm{V,R}}^{(n,n)}$ and $G_{\mathrm{M,R}}^{(0,n)}$ are the channel gains from VH n and the MBS to RU n, respectively.

If at least one of the three conditions above does not hold, while video i is cached in FBS j, the SINR at RU n obtained from FBS j is computed as

$$\gamma_{\mathrm{F,R}}^{(j,n,k,i)} = \frac{u_{j,i}(1 - s_{n,i} v_{k,n} p_{n,i}) P_{\mathrm{F}}^{(j)} G_{\mathrm{F,R}}^{(j,n)}}{\sigma_0}, \tag{6.15}$$

where $G_{\mathrm{F,R}}^{(j,n)}$ is the channel gain between FBS j and RU n.

Otherwise, if all the FBSs do not cache video i, the SINR at RU n is obtained from the MBS as

$$\gamma_{\mathrm{M,R}}^{(0,n,k,i)} = \frac{\prod_{j=1}^{J} (1 - u_{j,i})(1 - s_{n,i} v_{k,n} p_{n,i}) P_{\mathrm{M}}^{(0)} G_{\mathrm{M,R}}^{(0,n)}}{\sigma_0}. \tag{6.16}$$

The average capacity of all videos delivered to the RUs is expressed as

$$R_{\mathrm{R}} = W \left\{ \sum_{i=1}^{I} r_i \sum_{n=1}^{N} \sum_{k=1}^{K} \left[\log_2(1 + \gamma_{\mathrm{V,R}}^{(n,k,i)}) + \sum_{j=1}^{J} \log_2(1 + \gamma_{\mathrm{F,R}}^{(j,n,k,i)}) + \log_2(1 + \gamma_{\mathrm{M,R}}^{(0,n,k,i)}) \right] \right\}. \tag{6.17}$$

6.3.4 Objective Functions

In fact, there are two objective functions in the SRC optimization problem: one is for maximizing the cache-hit ratio and the other is for maximizing the overall average delivery capacity per MU. On the one hand, the objective of maximizing the cache-hit ratio is equivalent to maximizing the average number of caching copies $\overline{C}$ of all videos in J FBSs given by

$$\overline{C} = \sum_{i=1}^{I} r_i c_i, \tag{6.18}$$

where c_i is the number of copies of video i to be cached.

On the other hand, from (6.10), (6.13), and (6.17), the overall average delivery capacity per MU is given by

$$\overline{R} = \frac{R_{\mathrm{S}} + R_{\mathrm{T}} + R_{\mathrm{R}}}{K + 2N}. \tag{6.19}$$

6.4 SRC Optimization Design

The SRC optimization problem consists of the objective functions (6.18) and (6.19) and the two main constraints on (i) the storage capacity of the FBSs and (ii) the target SINR of the SHs. First, considering the storage capacity of the FBSs (C), the problem here is how to efficiently exploit C to maximize $\overline{C}$ for a high cache-hit ratio by finding c_i. This problem can be formulated as

$$\max_{c_i} \overline{C}, \tag{6.20}$$

$$s.t. \begin{cases} 0 \le c_i \le J, i = 1, 2, \ldots, I, \\ \sum_{i=1}^{I} c_i \le C, I \le C \le IJ. \end{cases} \tag{6.21}$$

The linear programming optimization problem above can be solved by using the primal-dual interior point method [17, 18].

Second, taking the target SINR (γ_0) of the SHs into account, we ignore the term $(1 - u_{j,i})$ in (6.8) and let $\gamma_{\mathrm{M,S}}^{(0,k,i)} \ge \gamma_0$. This means that increasing γ_0 guarantees the SHs higher SINR. Given γ_0 and c_i found by solving (6.20) and (6.21), the problem of maximizing $\overline{R}$ is formulated as follows:

$$\max_{u_{j,i}, v_{k,n}} \overline{R}, \tag{6.22}$$

$$s.t. \begin{cases} \sum_{j=1}^{J} u_{j,i} \le c_i, i = 1, 2, \ldots, I, \\ \sum_{n=1}^{N} s_{n,i} v_{k,n} p_{n,i} P_{\mathrm{V}}^{(n)} G_{\mathrm{V,S}}^{(n,k)} \le \dfrac{P_M^{(0)} G_{M,S}^{(0,k)}}{\gamma_0} - \sigma_0, \\ k = 1, 2, \ldots, K, i = 1, 2, \ldots, I. \end{cases} \tag{6.23}$$

In (6.22) and (6.23), finding the optimal results of $u_{j,i}$ and $v_{k,n}$ is equivalent to finding the two optimal matrices $U^*_{J\times I}$ and $V^*_{K\times N}$ in the two search spaces $\mathcal{U} = \{U^1_{J\times I}, U^2_{J\times I}, ..., U^{2^{J\times I}}_{J\times I}\}$ and $\mathcal{V} = \{V^1_{K\times N}, V^2_{K\times N}, ..., V^{2^{K\times N}}_{K\times N}\}$, respectively. For ease of implementation, we apply exhaustive search to find $U^*_{J\times I}$ and $V^*_{K\times N}$ [19] as in Algorithm 6.1. However, the memory and time complexities of Algorithm 6.1 make it impractical for implementation by the MBS in a large-scale network such as 6G UDN. Alternately, the MBS divides the total search space into multiple sub-search spaces and assigns the FBSs to exhaustively search in the separate sub-search spaces. Then, the FBSs send their suboptimal solutions to the MBS for finding the global optimal solution.

Algorithm 6.1 Exhaustive Search

Input: Initial parameters given in Table 6.1
Output: $\overline{R}^*$, $U^*_{J\times I}$, $V^*_{K\times N}$
1: Generate two feasible search spaces $\mathcal{U}' \in \mathcal{U}$ and $\mathcal{V}' \in \mathcal{V}$ that satisfy (6.23)
2: $\mathcal{R} \leftarrow \emptyset$
3: **for** each matrix $U_{J\times I}$ in $\mathcal{U}'$ **do**
4:     **for** each matrix $V_{K\times N}$ in $\mathcal{V}'$ **do**
5:         $\mathcal{R} \leftarrow \mathcal{R} \cup \overline{R}(U_{J\times I}, V_{K\times N})$, i.e., computing (6.19)
6:     **end for**
7: **end for**
8: $\overline{R}^* = \max \mathcal{R}$
9: $\{U^*_{J\times I}, V^*_{K\times N}\} = \arg\max \mathcal{R}$.

6.5 Performance Evaluation of SRC

6.5.1 System Setting

The system parameters are listed in Table 6.1. We compare the SRC with four other schemes: non-social-aware (NSA), non-downlink resource sharing (NRS), average performance (AVE), and minimum performance (MIN). In NSA, we change the characteristics of social relationships by generating all the permutations of $\{\kappa_n\}$ to have new encounter duration distributions with respect to (w.r.t) different permutations. Then, the worst permutation is selected for NSA. In NRS, no downlink resources are shared by the SHs. In AVE and MIN, the system delivery capacities are respectively averaged and minimized over $\mathcal{R}$ in Algorithm 6.1. It is noted that this chapter does not investigate the performance of high cache-hit ratio of SRC method, which will be analyzed in Chapter 7.

Table 6.1 Parameters setting.

Symbols	Specifications
$\{I, J, K, N\}$	$\{5, 3, 3, 5\}$
$\{T^{(i)}_{\min}\}$	$\{1, 15, 10, 5, 20\}$ s
$\{\theta_n\}$ and $\{\kappa_n\}$	$\{5, 10, 20, 15, 25\}$ and $\{1, 4, 3, 2, 5\}$ [9]
$\{P^{(0)}_M, P^{(j)}_F, P^{(n)}_T, \sigma_0\}$	$\{5$ W, 1 W, 1 mW, 10^{-13} W [20]$\}$
$\{\eta, \gamma_0, W\}$	$\{4$, 5 dB, 5 MHz$\}$
$\{\alpha, \delta, a, b\}$ and $\{\beta_n\}$	$\{2, 10, 0.5, 0.5\}$ and $\{0.1, 0.5, 0.9, 0.3, 0.7\}$
C	$0.7 \times IJ = 17.5$
Distances	**Specifications**
MBS to MUs	Randomly uniform distributed in [300, 1500] m
FBSs to MUs	Randomly uniform distributed in [50, 250] m
VHs to SHs	Randomly uniform distributed in [50, 100] m
VHs to RUs	Randomly uniform distributed in [1, 50] m

6.5.2 System Performance Versus C

We first evaluate the system performance versus the caching storage capacity of FBSs by setting $C = \{0, 0.1, 0.2, ..., 1\} \times IJ$ in which $C = IJ$ means full caching strategy, i.e., all videos are cached in all FBSs. As shown in Figure 6.2, it is not reasonable to deploy a full caching strategy because the performance does not increase significantly when C approaches IJ. A proper portion of caching storage capacity is $C = 0.7 \times IJ$ at the elbow of SRC method. Obviously, increasing C enhances the average system delivery capacity except for the MIN scheme. In comparison, the SRC always outperforms the others. The NSA and NRS are similar and better than the AVE and MIN, while the MIN is the worst one. The lower performance of NSA and NRS demonstrates the role of social-aware and downlink resource sharing in improving system performance.

6.5.3 System Performance Versus δ

Figure 6.3 plots the system performance versus different video durations by changing δ in (6.5) from 1 to 40. The results show that the duration of videos can be adjusted to meet the social relationships of D2D pairs for the highest system performance, i.e., at $\delta = 10$. The reason is that a low value of δ results in a high probability of successfully offloading the videos over D2D communications, which, however, imposes significant interference on the SHs. Meanwhile, a high value of δ results in a low probability of success and thus fails to efficiently utilize

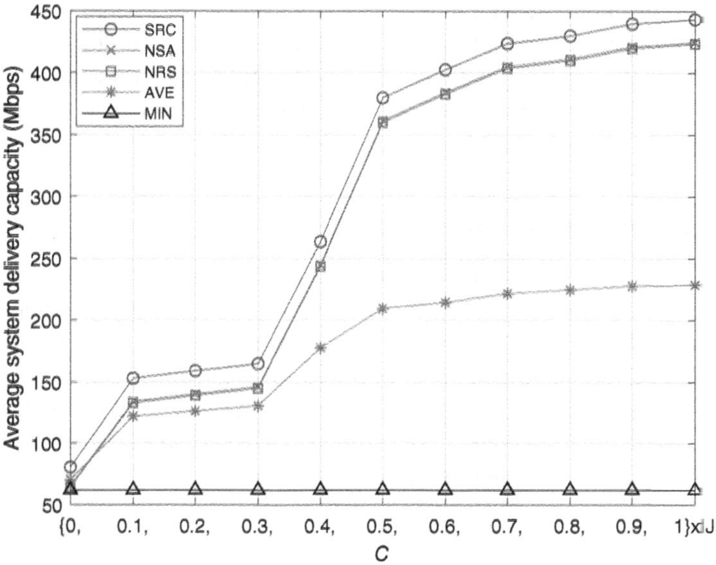

Figure 6.2 Capacity performance versus caching storage capacity of FBSs. *Source*: Reproduced from [1] / Springer Nature.

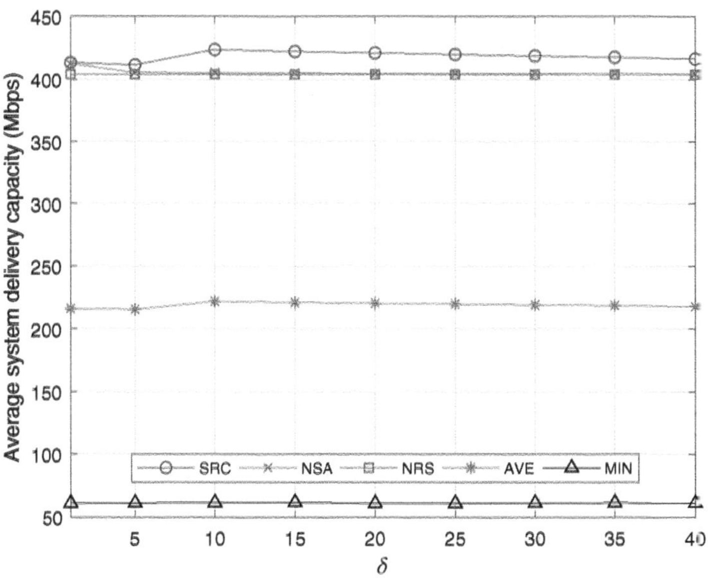

Figure 6.3 Capacity performance versus coefficient of video length adjustment. *Source*: Reproduced from [1] / Springer Nature.

D2D communications. Both low and high values of δ degrade the system performance. Obviously, the NRS is not affected by δ. In comparison, the performance of SRC is better than the others, and it reduces to that of NSA and NRS when δ is too low or too high. The performance of NSA and NRS is similar and better than that of AVE and MIN. The MIN is always the worst performance.

6.5.4 System Performance Versus α

In Figure 6.4, we evaluate the system performance versus the skewed popularity coefficient of videos (α). The results demonstrate that exploiting the skewed popularity can improve the system performance. If the videos have the same popularity ($\alpha = 0$), high performance gain is not achieved, whereas much higher gain can be obtained as α increases. In other words, focusing on serving a smaller number of videos with higher popularity certainly provides the MUs with greater performance gain. The SRC thus obtains the best system performance compared to others versus α.

6.5.5 System Performance Versus N

In Figure 6.5, the system performance is investigated by changing the number of D2D pairs (N). The parameters of the probability density function for encounter

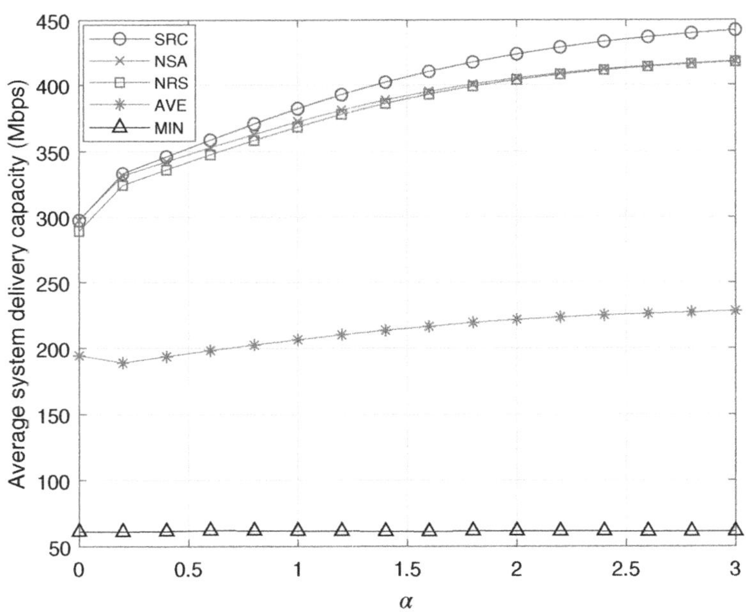

Figure 6.4 Capacity performance versus skewed access rate of videos.
Source: Reproduced from [1] / Springer Nature.

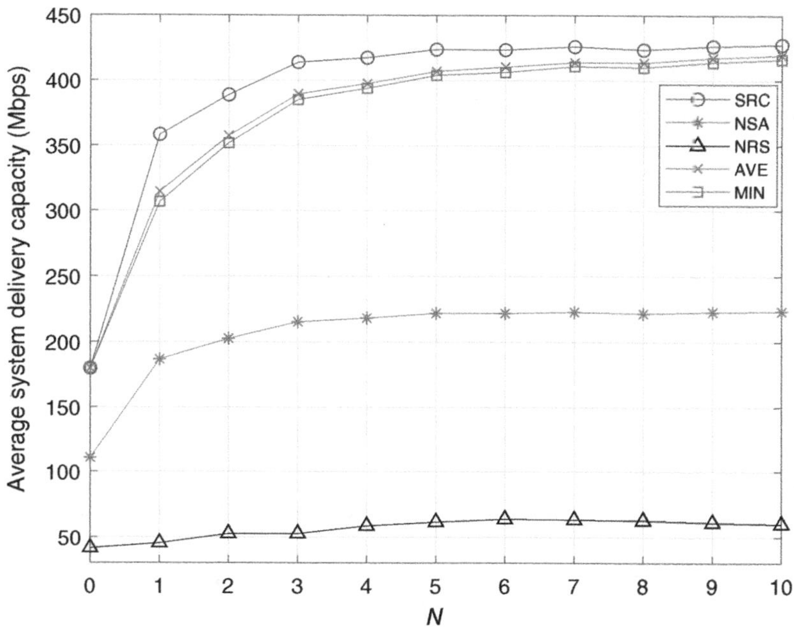

Figure 6.5 Capacity performance versus number of D2D pairs. *Source*: Reproduced from [1] / Springer Nature.

duration are also expanded to $\{\theta_n\} = \{5, 10, 20, 15, 25, 10, 5, 20, 25, 15\}$, $\{\kappa_n\} = \{1, 4, 3, 2, 5, 3, 2, 4, 1, 5\}$, and $\{\beta_n\} = \{0.1, 0.5, 0.9, 0.3, 0.7, 1, 0.2, 0.4, 0.6, 0.8\}$. It is easy to see that when $N = 0$, the system does not gain benefits from downlink resource sharing and social relationships of D2D pairs. Therefore, the SRC, NSA, and NRS have the same low result of system delivery capacity. When N increases, the system delivery capacity increases and then gets saturated. The saturated situation holds due to the constraint on the target SINR of SHs that limits the amount of interference generated by D2D communications. In the context of 6G UDN with a massive number of D2D pairs, to reduce system design complexity, a suitable value of N is carefully selected such that the target SINR of the SHs is guaranteed while ensuring sufficiently high system delivery capacity before reaching the saturated situation. The SRC gains the best performance compared to the others.

6.5.6 System Performance Versus J

Figure 6.6 plots the system performance versus the number of FBSs (J). The results show that the number of FBSs plays an important role in improving the system performance. Without considering the constraint on cross-tier and co-tier interferences by applying channel splitting and F-ALOHA schemes

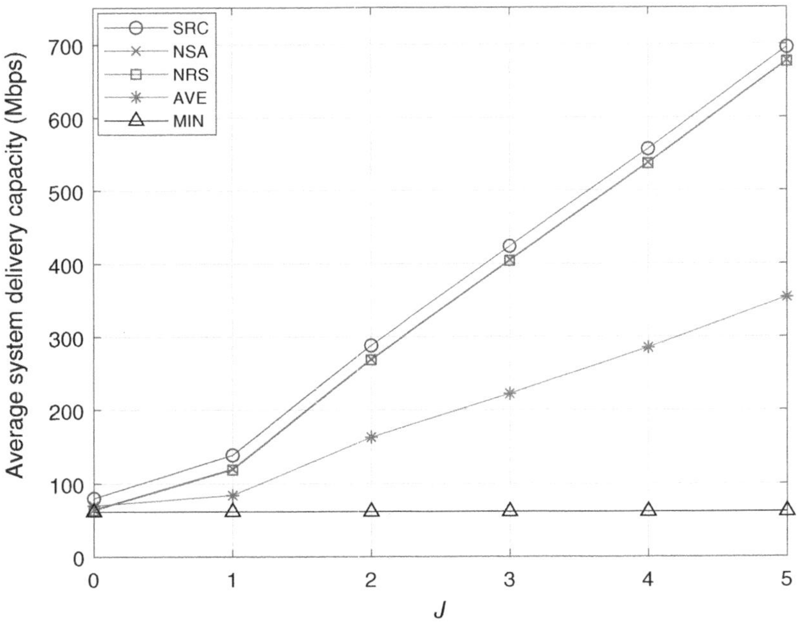

Figure 6.6 Capacity performance versus number of FBSs. *Source*: Reproduced from [1] / Springer Nature.

[15, 16], the system delivery capacity does not get a saturated situation w.r.t the increase of J. We can see that the SRC outperforms the others, but the performance gain between the SRC and the other two NSA and NRS does not change because they utilize the same optimal caching placements.

6.5.7 System Performance Versus γ_0

We investigate the system performance under the impact of the target SINR of the SHs (γ_0) as shown in Figure 6.7. When γ_0 increases, the system delivery capacity of SRC and NSA decreases to that of NRS. The reason is that if γ_0 is low, more D2D pairs share the downlink resource from the SHs to offload the videos for higher system delivery capacity. Otherwise, if γ_0 is high, fewer D2D pairs are established to offload the videos, which lowers the system delivery capacity. When γ_0 increases, the system delivery capacity of MIN increases because more candidate matrices that cause higher interference impact on the SHs are removed from the search space $\mathcal{V}$. It is certain that the result of NRS is not affected by γ_0 due to no interference effect from D2D communications on the SHs. Importantly, the proposed SRC always surpasses the other methods.

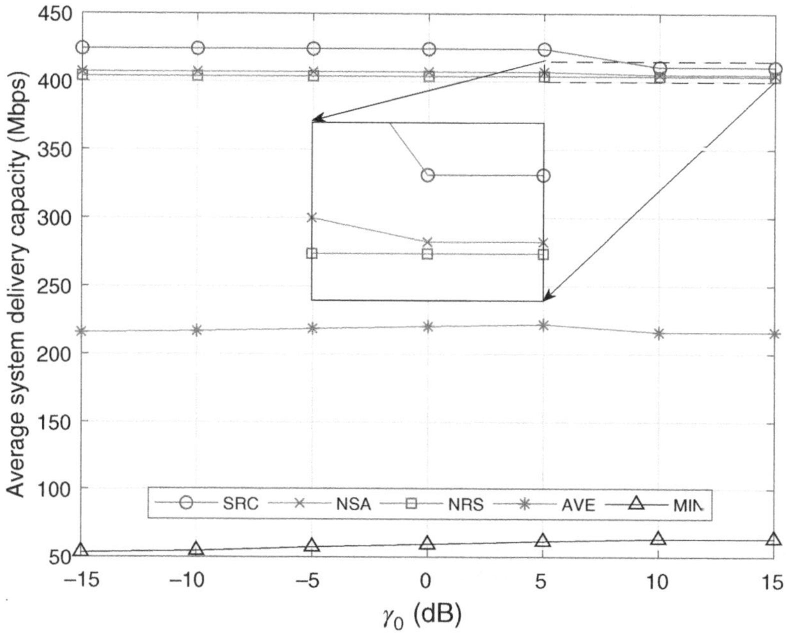

Figure 6.7 Capacity performance versus target SINR of SHs. *Source*: Reproduced from [1] / Springer Nature.

6.5.8 System Performance Versus Distance Between MBS and MUs

We further evaluate the system performance versus the distance between MBS and MUs in the range of $d_1 = [5, 10]$m, $d_2 = [10, 20]$m, $d_3 = [20, 100]$m, $d_4 = [100, 500]$m, $d_5 = [200, 1000]$m, $d_6 = [300, 1500]$m, $d_7 = [400, 2000]$m, and $d_8 = [500, 2500]$m. In this scenario, we only consider the SRC with different numbers of FBS J from 1 to 5. Figure 6.8 shows that if the number of FBSs is low (i.e., $J = 1$ and $J = 2$) and the MUs are closer to the MBS (d_1 and d_2), the MUs are more likely to be served by the MBS at low system delivery capacity. Otherwise, if the number of FBSs increases, the MUs are more likely to be served by the FBSs with greater system delivery capacity. Thanks to the high caching storage capacity of the FBSs ($C = 0.7 \times IJ$), the MBS is not serving the MUs even though they are placed very close to it. Interestingly, there exists the best region (i.e., $d_6 = [300, 1500]$m) in which the MUs are served with the highest system delivery capacity. The optimal region is neither too close to nor too far from the MBS. In other words, the MUs in this region are optimally served by all the MBS, the FBSs, and the VHs.

Finally, we consider the lower caching storage capacity of the FBSs by setting $C = \{0.6, 0.5, 0.4, 0.3\} \times IJ$. Figure 6.9 shows that when C is too low ($C = 0.3 \times IJ$),

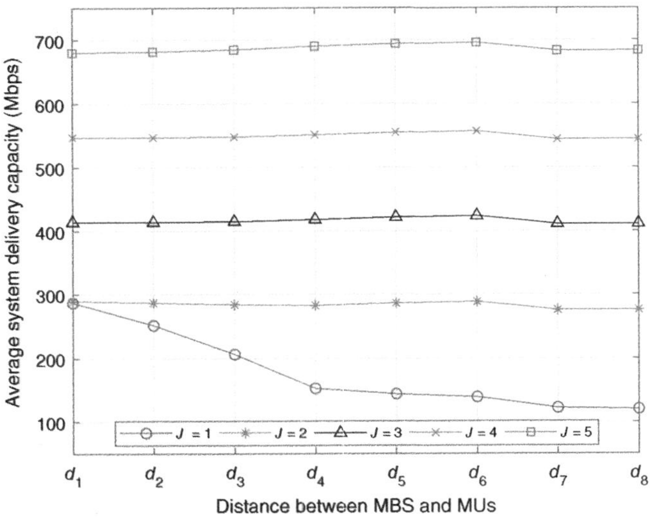

Figure 6.8 Capacity performance versus the distance between MBS and MUs. *Source*: Reproduced from [1] / Springer Nature.

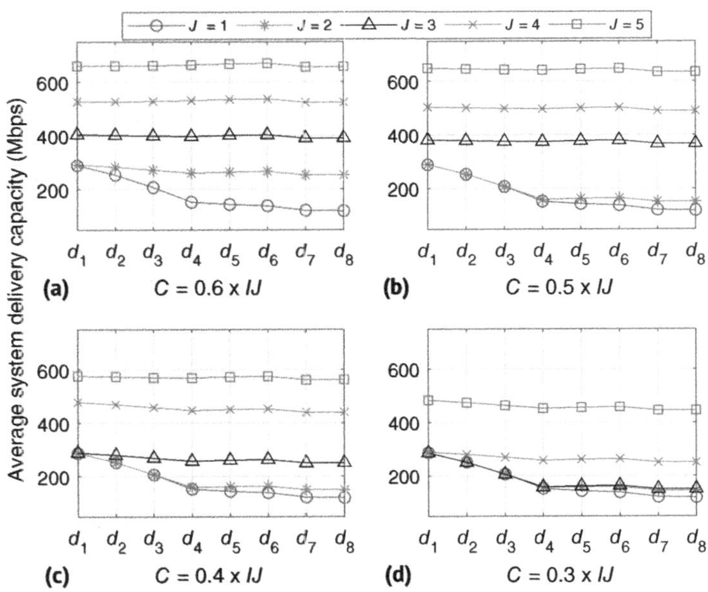

Figure 6.9 Capacity performance versus the distance between MBS and MUs with lower C. *Source*: Reproduced from [1] / Springer Nature.

the MUs are mainly served by the MBS when they are closer to the MBS. Especially when $d_1 = [5, 10]$m, the performance of the SRC is the same in all cases of $J = 1$, $J = 2$, $J = 3$, and $J = 4$, but the FBSs actually join in to serve the MUs when $J = 5$. When C and J increase, the MUs are more likely to be served by the FBSs than by the MBS. Hence, increasing both C and J can significantly mitigate the workload of the MBS in 6G UDN.

6.6 Conclusion

We have proposed a SRC method for cooperative video transmission in 6G UDN. In particular, the storage resource of the MBS, FBSs, and VHs and the downlink spectrum resource of the SHs were utilized to provide a more flexible access to VAS over a shorter transmission distance. For more efficiency, we exploited the social relationships of D2D pairs and the popularity of videos. The SRC method also considered the target SINR of the SHs to guarantee their QoS. As a result, the proposed SRC method could (i) relax the workload of the MBS and FBSs and (ii) provide the MUs with a high cache-hit ratio and a high delivery capacity of VAS alternately requested from the MBS, FBSs, and VHs. It also introduced two findings: (i) a proper duration set of videos in accordance with the social relationships of D2D pairs that can provide the MUs with the highest delivery capacity, and (ii) there exists the best region in which the MUs are served with the highest delivery capacity. These findings can efficiently assist Internet service and content providers in designing VAS over 6G UDN.

References

1 M.-P. Bui, N.-S. Vo, S. Q. Nguyen, and Q.-N. Tran. Social-aware caching and resource sharing maximized video delivery capacity in 5G ultra-dense networks. *Springer Mobile Netw. Appl.*, 25:2037–2049. 2019.

2 Z. Wang, J. Liu, and W. Zhu. Social-aware video delivery: Challenges, approaches, and directions. *IEEE Netw.*, 30(5):35–39. 2016.

3 D. Wu, J. Li, P. He, Y. Cui, and R. Wang. Social-aware graph-based collaborative caching in edge-user networks. *IEEE Trans. Veh. Technol.*, 72(6):7926–7941. 2023.

4 Z. Su, Q. Xu, F. Hou, Q. Yang, and Q. Qi. Edge caching for layered video contents in mobile social networks. *IEEE Trans. Multimed.*, 19(10):2210–2221. 2017.

5 J. Cai, X. Wu, Y. Liu, J. Luo, and L. Liao. Network coding-based socially-aware caching strategy in D2D. *IEEE Access*, 8:12784–12795. 2020

6 R. Zhang, S. Jia, Y. Ma, and C. Xu. Social-aware D2D video delivery method based on mobility similarity measurement in 5G ultra-dense network. *IEEE Access*, 8:52413–52427. 2020.

7 Y. Chiang, C.-H. Hsu, and H.-Y. Wei. Collaborative social-aware and QoE-driven video caching and adaptation in edge network. *IEEE Trans. Multimed.*, 23(0):4311–4325. 2020.

8 Y. Li, G. Wu, and J. Tang. Social-aware edge caching for UAV-assisted metaverse systems. In *Proc. IEEE Global Commun. Conf.*, pages 1–6, Kuala Lumpur, Malaysia, Dec. 2023.

9 X. Zhang, Y. Li, Y. Zhang, J. Zhang, H. Li, S. Wang, and D. Wang. Information caching strategy for cyber social computing based wireless networks. *IEEE Trans. Emerg. Top. Comput.*, 5(3):391–402. 2017.

10 Y. Zhang, E. Pan, L. Song, W. Saad, Z. Dawy, and Z. Han. Social network aware device-to-device communication in wireless networks. *IEEE Trans. Wirel. Commun.*, 14(1):177–190. 2015.

11 J. Zhang, X. Zhang, Z. Yan, Y. Li, W. Wang, and Y. Zhang. Social-aware cache information processing for 5G ultra-dense networks. In *Proc. Int. Conf. Wireless Commun. and Signal Processing*, pages 1–5, Yangzhou, China, Nov. 2016.

12 M. Cha, H. Kwak, P. Rodriguez, Y.-Y. Ahn, and S. Moon. I tube, you tube, everybody tubes: Analyzing the world's largest user generated content video system. In *Proc. ACM SIGCOMM Inter. Measurement Conf.*, pages 1–14, New York, USA, Oct. 2007.

13 F. Benevenuto, T. Rodrigues, M. Cha, and V. Almeida. Characterizing user behavior in online social networks. In *Proc. ACM SIGCOMM Inter. Measurement Conf.*, pages 49–62, New York, USA, Nov. 2009.

14 B. Bai, L. Wang, Z. Han, W. Chen, and T. Svensson. Caching based socially-aware D2D communications in wireless content delivery networks: A hypergraph framework. *IEEE Wirel. Commun.*, 23(4):74–81. 2016.

15 V. Chandrasekhar and J. G. Andrews. Spectrum allocation in tiered cellular networks. *IEEE Trans. Commun.*, 57(10):3059–3068. 2009.

16 W. C. Cheung, T. Q. S. Quek, and M. Kountouris. Throughput optimization, spectrum allocation, and access control in two-tier femtocell networks. *IEEE J. Sel. Areas Commun.*, 30(3):561–574. 2012.

17 S. Mehrotra. On the implementation of a primal-dual interior point method. *SIAM J. Optim.*, 2:575–601. 1992.

18 Y. Zhang. Solving large-scale linear programs by interior-point methods under the MATLAB environment. *Optim. Methods Softw.*, 10(1):1–31. 2007.

19 N.-S. Vo, T. Q. Duong, M. Guizani, and A. Kortun. 5G optimized caching and downlink resource sharing for smart cities. *IEEE Access*, 6:31457–31468. 2018.

20 N. Eshraghi, V. Shah-Mansouri, and B. Maham. QoE-aware power allocation for device-to-device video transmissions. In *Proc. IEEE Personal, Indoor and Mobile Radio Commun.*, pages 1–5, Valencia, Spain, Sep. 2016.

7

Spectrum Sharing and Caching for Multimedia Applications and Services in Content Delivery Networks

7.1 Introduction

In smart cities, a multilayer platform of information, communications, and sensing technologies based CDNs is developed to capture the surrounding environments and events for the purpose of monitoring, analyzing, decision-making, management, and further research. Among various types of data (e.g., from sensors, mobile and wearable devices, cameras, actuators, and machines) are multimedia contents of surveillance, management, e-learning and e-health, and infotainment applications and services.[1]

Ericsson estimates that in 2029, there will be 39 billion IoT devices connected to 6G networks, which will certainly generate a tremendous amount of multimedia traffic. It is a challenge to distribute multimedia contents to a massive number of citizens in smart cities or MUs in 6G networks by conventional communications from MBSs. To address this, the idea of carrying the resources and capabilities for sensing, communications, computing, storage, and intelligence (SCCSI) with vehicles [2, 3] is proposed. In this way, various advanced MAS can be pushed to the edge of CDN by using a vehicle as a service (VaaS) with SCCSI networks. Importantly, the authors in [4] have explored the potential integration of artificial intelligence (AI) and digital twin in 6G networks and discussed their practical deployment in smart cities by utilizing the resource allocation, caching, offloading, and security approaches.

In advanced applications and services of a smart city, the integration of wireless sensor networks enabled IoT and mobile cellular networks has been studied over the years, from 5G to 6G, as a feasible architecture for collecting sensed data [5–8]. In terms of mobile and multi-access edge computing, edge caching plays an important role in providing emerging IoT applications and services with high cache hit ratio [9]. Intelligent caching, user association, and information-centric cognitive

1 This chapter has been published partly in [1].

radio networks are also promising solutions for reducing latency and improving connectivity, content hit ratio, and reliability, particularly for MAS such as high-definition video, online gaming, and virtual/augmented reality in smart cities [10, 11].

In this chapter, to provide the MUs with MAS over CDN in smart cities, we extend our previous work [12] by taking into account the integration of M-IoT and 6G networks in the context of smart cities. Particularly, a downlink spectrum SCO solution is deployed in the same manner as in Chapter 6. However, MAS in smart cities faces a serious congestion challenge due to a huge amount of rich data generated by M-IoT devices and then requested by a massive number of MUs in 6G networks. To address this challenge, we further focus on maximizing the average number of replicas of contents, i.e., NRO solution, so that the MUs can access the MAS at high hit rates besides being served with high system delivery capacity. In other words, the number of replicas to cache is found in accordance with the popularity and the demand of multimedia contents to prevent wasted throughput, e.g., reserved throughput is much larger than requested throughput. Consequently, we design a joint downlink spectrum SCF consisting of two optimization solutions: NRO and SCO.

The rest of this chapter is organized as follows. In Section 7.2, we introduce the system model of SCF in 6G networks with integrated M-IoT. The system is analyzed to derive the hit rate and system delivery capacity in Section 7.3. Section 7.4 presents the NRO and SCO problems and solutions. Section 7.5 demonstrates the benefits of the proposed SCF compared with other schemes through simulation results. Finally, Section 7.6 concludes the chapter.

7.2 System Model of SCF

The SCF model for CDN over 6G networks with integrated M-IoT is illustrated in Figure 7.1. An MBS, J femto-cell base stations (FBSs), K spectrum helping MUs (SHs), and N D2D pairs of video helping MUs (VHs) and requesting MUs (RUs) are deployed in the same manner as in Chapter 6 but without social relationships considered. The M-IoT devices are grouped into I clusters, each with a cluster head (CH) responsible for covering a specific area and capturing multimedia content through a joint active duty scheduling and encoding rate allocation scheme [6]. CH i is connected to the MBS or to an arbitrary FBS via a link with capacity R_i Mbps.

Normally, all the contents captured by M-IoT devices are sent to the MBS. The MBS can either cache the multimedia contents to serve the MUs if requested or forward them to the data center for monitoring, analyzing, decision-making, management, and further research. To mitigate the MUs' requests to the MBS, the

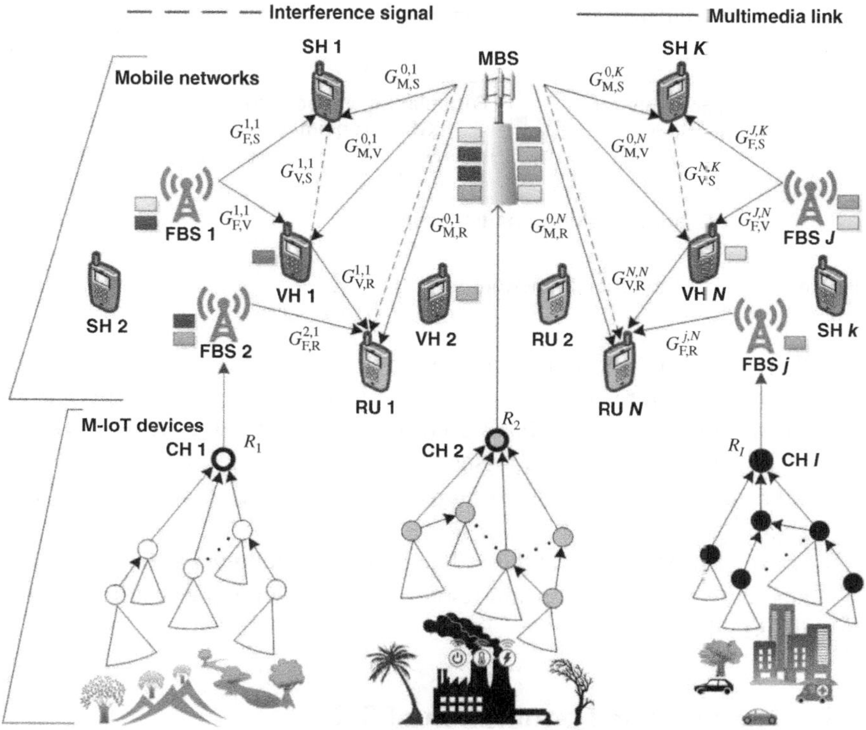

Figure 7.1 SCF model for CDN.

multimedia contents are cached in the FBSs. In addition, the multimedia contents are also offloaded from the VHs who have already cached them before. Caching in the FBSs is deployed to provide a high hit ratio to access the multimedia contents and enhance the system delivery capacity, whereas caching in the VHs mainly aims to improve the system delivery capacity.

In this model, the multimedia contents can be streamed from (i) the MBS and FBSs to the SHs and VHs and (ii) the MBS, FBSs, and VHs to the RUs, thanks to multiple antennas at each MU [13, 14]. The detailed SCF, which is executed by the MBS for high hit rate and high system delivery capacity, is shown in Figure 7.2 and works as follows:

- Step 1: The MBS updates all the system parameters if there are any significant changes.
- Step 2: Based on the system parameters, the MBS formulates and solves the NRO problem for high hit ratio by finding the optimal number of replicas of content i (c_i), $i = 1, 2, ..., I$.

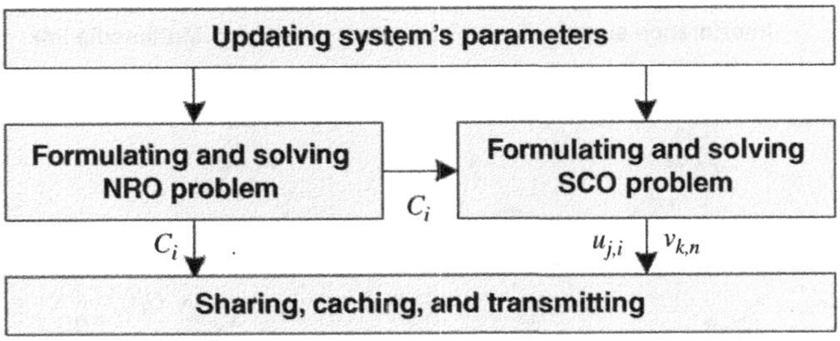

Figure 7.2 SCF optimization framework.

- Step 3: Based on c_i found in step 2, the MBS formulates and solves the CSO problem for maximum system delivery capacity by finding the optimal caching index $u_{j,i}$ and optimal spectrum sharing index $\upsilon_{k,n}$. Here $u_{j,i} = 1$ and $\upsilon_{k,n} = 1$ mean that FSB j is assigned to cache content i and SH k agrees to share its downlink spectrum resource with D2D pair n, otherwise $u_{j,i} = 0$ and $\upsilon_{k,n} = 0$.
- Step 4: The MBS deploys the SCF to cache the multimedia contents and to share the downlink spectrum resources for delivering the multimedia contents to the MUs.

7.3 Analysis and Formulation of SCF

7.3.1 Hit Ratio

A high hit ratio can be ensured by placing the number of replicas c_i of content i proportionally to its popularity r_i [15]. To improve hit ratio performance in NRO problem, we maximize the average number of replicas, which is given by

$$\overline{I} = \sum_{i=1}^{I} r_i c_i. \tag{7.1}$$

7.3.2 System Delivery Capacity

Given the channel models in Chapter 6 [16, 17], the system delivery capacity from the MBS, FBSs, and VHs to the MUs is derived in the sequel.

SH k can share its downlink resource with D2D pair n while requesting the multimedia contents from the MBS and FBSs. The signal-to-interference-plus-noise ratios (SINRs) of the channels from the MBS and FBSs to SH k are given by

$$\gamma_{\mathrm{M,S}}^{(0,k,i)} = \frac{P_{\mathrm{M}}^{(0)} G_{\mathrm{M,S}}^{(0,k)}}{\sigma_0 + \sum_{n=1}^{N} \upsilon_{k,n} p_{n,i} P_{\mathrm{V}}^{(n)} G_{\mathrm{V,S}}^{(n,k)}}, \tag{7.2}$$

$$\gamma_{\mathrm{F,S}}^{(j,k,i)} = \frac{u_{j,i} P_{\mathrm{F}}^{(j)} G_{\mathrm{F,S}}^{(j,k)}}{\sigma_0}, \tag{7.3}$$

where $P_{\mathrm{M}}^{(0)}$ and $P_{\mathrm{F}}^{(j)}$ are the transmission powers of the MBS and FBS j, and $G_{\mathrm{M,S}}^{(0,k)}$ and $G_{\mathrm{F,S}}^{(j,k)}$ are the channel gains from the MBS and FBS j to SH k. σ_0 is the power of additive white Gaussian noise (AWGN). It is noted that SH k is affected by the interference from VH n of D2D pair in which $P_{\mathrm{V}}^{(n)}$ is the transmission power over the channel gain $G_{\mathrm{V,S}}^{(n,k)}$ between VH n and SH k. And $p_{n,i}$ is the probability that VH n has cached content i as defined in (6.9) of Chapter 6.

Similarly, the SINRs of the channels from the MBS and FBSs to VH n are

$$\gamma_{\mathrm{M,V}}^{(0,n)} = \frac{P_{\mathrm{M}}^{(0)} G_{\mathrm{M,V}}^{(0,n)}}{\sigma_0}, \tag{7.4}$$

$$\gamma_{\mathrm{F,V}}^{(j,n,i)} = \frac{u_{j,i} P_{\mathrm{F}}^{(j)} G_{\mathrm{F,V}}^{(j,n)}}{\sigma_0}. \tag{7.5}$$

The RUs of D2D pairs are further served by the VHs beside the MBS and FBSs. The SINRs of the channels from the MBS, FBS j, and VH n to RU n are computed as

$$\gamma_{\mathrm{M,R}}^{(0,n)} = \frac{P_{\mathrm{M}}^{(0)} G_{\mathrm{M,R}}^{(0,n)}}{\sigma_0}, \tag{7.6}$$

$$\gamma_{\mathrm{F,R}}^{(j,n,i)} = \frac{u_{j,i} P_{\mathrm{F}}^{(j)} G_{\mathrm{F,R}}^{(j,n)}}{\sigma_0}, \tag{7.7}$$

$$\gamma_{\mathrm{V,R}}^{(n,k,i)} = \frac{v_{k,n} p_{n,i} P_{\mathrm{V}}^{(n)} G_{\mathrm{V,R}}^{(n,n)}}{\sigma_0 + P_{\mathrm{M}}^{(0)} G_{\mathrm{M,R}}^{(0,n)} + \sum_{l=1,l\neq n}^{N} v_{k,l} p_{l,i} P_{\mathrm{V}}^{(l)} G_{\mathrm{V,R}}^{(l,l)}}, \tag{7.8}$$

In (7.8), RU n is affected by the interference from both the MBS with transmission power $P_{\mathrm{M}}^{(0)}$ and the other VH l with transmission power $P_{\mathrm{V}}^{(l)}$, $l \neq n$, when reusing the same spectrum resource shared by SH k.

From (7.2) to (7.8), the capacities delivered to SHs, VHs, and RUs are, respectively, given by

$$R_{\mathrm{S}} = W \sum_{k=1}^{K} \sum_{i=1}^{I} r_i \left[\log_2 \left(1 + \gamma_{\mathrm{B,S}}^{(0,k,i)}\right) + \sum_{j=1}^{J} \log_2(1 + \gamma_{\mathrm{F,S}}^{(j,k,i)}) \right], \tag{7.9}$$

$$R_{\mathrm{V}} = W \sum_{n=1}^{N} \left[\log_2 \left(1 + \gamma_{\mathrm{M,V}}^{(0,n)}\right) + \sum_{j=1}^{J} \sum_{i=1}^{I} r_i \log_2(1 + \gamma_{\mathrm{F,V}}^{(j,n,i)}) \right], \tag{7.10}$$

$$R_{\mathrm{R}} = W\sum_{n=1}^{N}\left[\log_2(1+\gamma_{\mathrm{M,R}}^{(0,n)}) + \sum_{j=1}^{J}\sum_{i=1}^{I} r_i \log_2(1+\gamma_{\mathrm{F,R}}^{(j,n,i)}) + \sum_{k=1}^{K}\sum_{i=1}^{I} r_i \log_2(1+\gamma_{\mathrm{V,R}}^{(n,k,i)})\right]. \quad (7.11)$$

Finally, the overall average system capacity delivered to each MU (i.e., SH, VH, and RU) is expressed as

$$R = \frac{R_{\mathrm{S}} + R_{\mathrm{V}} + R_{\mathrm{R}}}{K + 2N}. \quad (7.12)$$

The above overall average system capacity can be maximized by finding the optimal caching index $u_{j,i}$ and optimal sharing index $v_{k,n}$ in the SCO problem, which is formulated in Section 7.4.

7.4 SCF Optimization Design

7.4.1 NRO

As in (7.1), a high hit ratio can be obtained by finding the optimal number of replicas of each content c_i to maximize the average number of replicas $\bar{I}$. To do so, we further take into account the constraints on (i) caching capacity of FBSs and (ii) reserved throughput of MBS, FBSs, and VHs ($R_{\mathrm{Res}}^{(i)}$) versus required throughput of all MUs ($R_{\mathrm{Req}}^{(i)}$). The NRO problem is formulated as

$$\max_{c_i} \bar{I}, \quad (7.13)$$

$$\text{s.t.} \begin{cases} 1 \le c_i \le J, & i = 1, 2, \dots, I, \\ \sum_{i=1}^{I} c_i \le \rho I J, & \frac{1}{J} \le \rho \le 1, \\ R_{\mathrm{Res}}^{(i)} \le R_{\mathrm{Req}}^{(i)}, & i = 1, 2, \dots, I, \end{cases} \quad (7.14)$$

where the first and second constraints are used to ensure that at least one replica of each content and the upper amount of replicas of all contents are cached in the FBSs. The third constraint is to avoid the wasted throughput served by the MBS, FBSs, and VHs. The $R_{\mathrm{Res}}^{(i)}$ and $R_{\mathrm{Req}}^{(i)}$ for the i-th multimedia content are respectively given by

$$R_{\mathrm{Res}}^{(i)} = R_{\mathrm{M}}^{(i)} + R_{\mathrm{F}}^{(i)} c_i + R_{\mathrm{V}}^{(i)} \sum_{n=1}^{N} p_{n,i}, \quad (7.15)$$

$$R_{\text{Req}}^{(i)} = (K + 2N)R_i r_i, \tag{7.16}$$

where $R_{\text{M}}^{(i)}$, $R_{\text{F}}^{(i)}$, and $R_{\text{V}}^{(i)}$ are the reserved throughput at the MBS, FBSs, and VHs for content i to serve the MUs.

By substituting (7.15) and (7.16) for the third constraint of (7.14), we have

$$c_i \leq \frac{R_{\text{Req}}^{(i)} - (R_{\text{V}}^{(i)} \sum_{n=1}^{N} p_{n,i} + R_{\text{M}}^{(i)})}{R_{\text{F}}^{(i)}}. \tag{7.17}$$

Combining (7.17) and the first constraint of (7.14), the NRO problem can be rewritten as

$$\max_{c_i} \bar{I} \tag{7.18}$$

$$\text{s.t.} \begin{cases} 1 \leq c_i \leq C_i, & i = 1, 2, \ldots, I, \\ \sum_{i=1}^{I} c_i \leq \rho I J, & \frac{1}{J} \leq \rho \leq 1, \end{cases} \tag{7.19}$$

where

$$C_i = \min \left\{ J, \max\{1, \frac{R_{\text{Req}}^{(i)} - (R_{\text{V}}^{(i)} \sum_{n=1}^{N} p_{n,i} + R_{\text{M}}^{(i)})}{R_{\text{F}}^{(i)}}\} \right\}. \tag{7.20}$$

This is a linear programming optimization problem and can be solved by using a primal-dual interior point method (a variant of Mehrotra's predictor–corrector algorithm) [18, 19]. The operator max in (7.20) is to guarantee that the upper bound C_i of the first constraint in (7.19) cannot be less than 1.

7.4.2 Sharing and Caching Optimization

For the SCO problem, taking into account the constraints on (i) c_i found by solving (7.13) and (7.14) and (ii) the target SINR γ_0 of SHs, the SCO problem, which is formulated and solved to maximize the overall system delivery capacity R in (7.12) by finding $u_{j,i}$ and $v_{k,n}$, is given by

$$\max_{u_{j,i}, v_{k,n}} R \tag{7.21}$$

$$
\text{s.t.} \begin{cases} \sum_{j=1}^{J} u_{j,i} \leq c_i, i = 1, 2, \ldots, I, \\ \sum_{n=1}^{N} \upsilon_{k,n} p_{n,i} P_{\mathrm{V}}^{(n)} G_{\mathrm{V,S}}^{(n,k)} \leq \dfrac{P_{\mathrm{M}}^{(0)} G_{\mathrm{M,S}}^{(0,k)}}{\gamma_0} - \sigma_0, \\ k = 1, 2, \ldots, K, i = 1, 2, \ldots, I, \end{cases} \tag{7.22}
$$

In (7.22), the first constraint is to limit the FBSs to caching content i to avoid wasted throughput. The second constraint comes from (7.2) by letting $\gamma_{\mathrm{M,S}}^{(0,k,i)} \geq \gamma_0$. It is used to limit the effect of interference of D2D pairs on the SHs. Finding the optimal results of $u_{j,i}$ and $\upsilon_{k,n}$ in (7.21) and (7.22) is equivalent to finding two optimal matrices $\mathbf{u}_{J\times I}^*$ and $\mathbf{v}_{K\times N}^*$ in the two matrix search spaces: $\mathcal{U} = \{u_{J\times I}^1, u_{J\times I}^2, \ldots, u_{J\times I}^{2^{J\times I}}\}$ and $\mathcal{V} = \{\upsilon_{K\times N}^1, \upsilon_{K\times N}^2, \ldots, \upsilon_{K\times N}^{2^{K\times N}}\}$, respectively. Exhaustive binary matrix search, which can be used to solve (7.21) and (7.22), is given in Algorithm 7.1.

Algorithm 7.1: Exhaustive Binary Search for SCO Problem

Input: Initial parameters given in Table 7.1
Output: $R^*, \mathbf{u}_{J\times I}^*, \mathbf{v}_{K\times N}^*$

```
Generating two matrix search spaces
  U = {u^1_{J×I}, u^2_{J×I}, ..., u^{2^{J×I}}_{J×I}} and
  V = {υ^1_{K×N}, υ^2_{K×N}, ..., υ^{2^{K×N}}_{K×N}}
R ← ∅
for each matrix u_{J×I} in U do
    for each matrix υ_{K×N} in V do
        if (7.22) satisfies then
            R(u_{J×I}, υ_{K×N}) = R, computing (7.12)
            R ← R ∪ R(u_{J×I}, υ_{K×N})
        end if
    end for
end for
R* = max R
{u*_{J×I}, v*_{K×N}} = argmax R
```

In Algorithm 7.1, the memory and time complexities, which depend on the total search spaces of $\mathcal{U}$ and $\mathcal{V}$, are equivalent to $\mathcal{O}(2^{J\times I + K\times N})$. In dense 6G networks, the SCO problem is difficult to be solved by centralized search at the MBS. In this scenario, the whole search space is divided into multiple sub-search spaces, and then distributed exhaustive binary search of each sub-search space can be independently done by each FBS. Afterwards, the sub-optimal results from the FBSs are sent to the MBS for finding global optimal solution.

7.5 Performance Evaluation of SCF

7.5.1 System Setting

We deploy the system with important parameters listed in Table 7.1. To evaluate the hit ratio performance of NRO, we compare it to the other two schemes named equal number of replicas (ENR) and worst number of replicas (WNR). In ENR, the number of replicas of each content $c_{\text{ENR}}^{(i)} = \frac{\sum_{i=1}^{I} c_i}{I}$, while in WNR, $c_{\text{WNR}}^{(i)}$ is inversely proportional to r_i such that $\sum_{i=1}^{I} c_{\text{WNR}}^{(i)} = \sum_{i=1}^{I} c_i$. To evaluate the system delivery capacity of SCO, we compare it to (i) average delivery capacity (ADC), (ii) worst delivery capacity (WDC), maximum delivery capacity without D2D offloading (NDO) [20], and (iii) maximum delivery capacity without femtocell caching (NFC) [21]. In ADC and WDC, the system delivery capacity is respectively averaged and minimized over the feasible combinations of the total search space. Meanwhile, in NDO and NFC, the system delivery capacity is maximized over two scenarios: (i) no caching in the VHs nor downlink resource sharing for D2D offloading and (ii) no caching in the FBSs.

7.5.2 Performance Metrics

7.5.2.1 Hit Ratio Performance

We first evaluate the performance of NRO, ENR, and WNR versus the total caching capacity of the FBSs by changing ρ in the range from 0.2 to 1 (7.19). In Figure 7.3, thanks to fully exploiting the popularity pattern of contents, NRO achieves the highest average number of replicas for the highest hit ratio compared to ENR and WNR. The higher caching capacity of FBSs increases the hit ratio but gets saturated if each FBS caches all the considered contents.

Figure 7.4 plots the hit ratio performance versus α in the range from 0 to 2. While the ENR does not change and the WNR decreases with respect to the increase of α, the average number of replicas of NRO increases to gain the best hit ratio performance until being saturated. In addition, considering the hit ratio performance versus J and I as illustrated in Figures 7.5 and 7.6, the proposed NRO also outperforms the ENR and WNR. The benefit of NRO can be achieved as the number of replicas c_i is proportional to the access rate r_i of content i, meanwhile the ENR and WNR cannot. In Figures 7.4 and 7.6, the hit ratio of WNR degrades versus the increase of α and I because $c_{\text{WNR}}^{(i)}$ is inversely proportional to r_i.

7.5.2.2 System Delivery Capacity Performance

For the ease of implementing the SCO, ADC, WDC, NDO [20], and NFC, we relax the memory and time complexities of searching by reducing the number of

Table 7.1 Parameters setting.

Symbols	Specifications
I	10 contents
J	5 FBSs
K	5 SHs
N	4 D2D pairs
$\{R_i\}$	{200, 190, 180, 170, 160, 150, 140, 130, 120, 110} Mbps
$\{R_{\mathrm{M}}^{(i)}\}$	{90, 80, 70, 60, 50, 40, 30, 20, 10, 5} Mbps
$\{R_{\mathrm{F}}^{(i)}\}$	{50, 45, 40, 35, 30, 25, 20, 15, 10, 5} Mbps
$\{R_{\mathrm{V}}^{(i)}\}$	{10, 9, 8, 7, 6, 5, 4, 3, 2, 1} Mbps
W	5 MHz
$P_{\mathrm{M}}^{(0)}$	10 W
$P_{\mathrm{F}}^{(j)}$	Fixed to 5 W
$\{P_{\mathrm{V}}^{(n)}\}$	{0.01, 0.05, 0.075, 0.1} W
γ_0	10 dB
σ_0	10^{-13} W
η	4 (path loss exponent)
α	1
ρ	0.5
$\{\beta_n\}$	{0.2, 0.4, 0.6, 0.8}
$\{a, b\}$	{0.5, 0.5}
Distances	**Specifications**
MBS to MUs	Randomly uniform distributed in [100, 500] m
FBSs to MUs	Randomly uniform distributed in [50, 250] m
VHs to SHs	Randomly uniform distributed in [50, 100] m
VHs to RUs	Randomly uniform distributed in [1, 50] m

contents from 10 to 4. The performance of system delivery capacity versus the total caching capacity of the FBSs is first evaluated by changing ρ as shown in Figure 7.7. We do not compare CSO to NFC because changing ρ is meaningless due to $J = 0$. We can see that the system delivery capacity of SCO, ADC, WDC, and NDO gets higher and then gradually saturates when ρ increases to cache more in the FBSs. In addition, without D2D offloading, the result of NDO is lower than the SCO, but it outpaces the ADC and WDC.

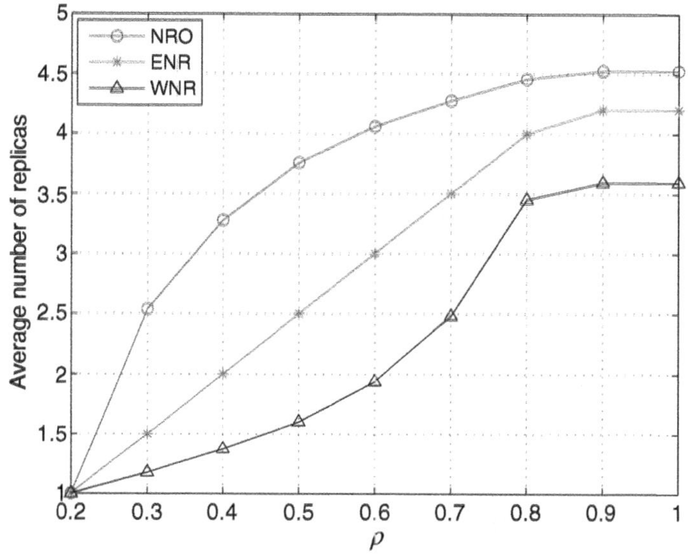

Figure 7.3 Hit ratio performance versus FBSs' caching capacity coefficient ρ.
Source: [1] / IEEE / CC BY 4.0.

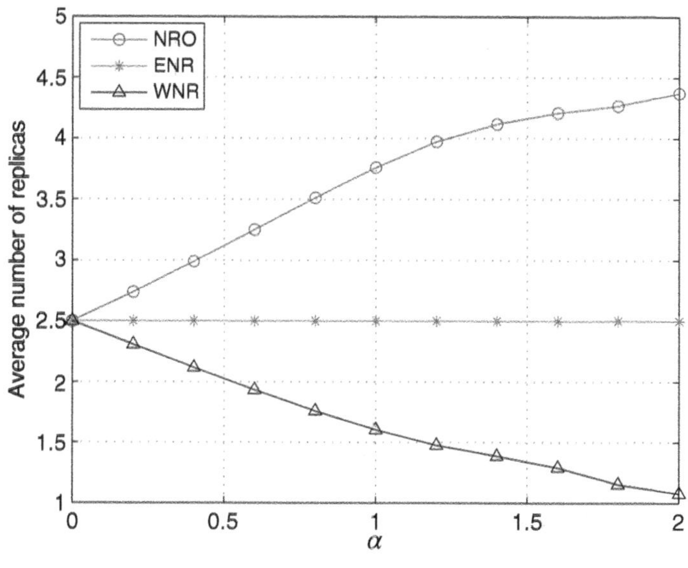

Figure 7.4 Hit ratio performance versus skewed access rate coefficient α.
Source: [1] / IEEE / CC BY 4.0.

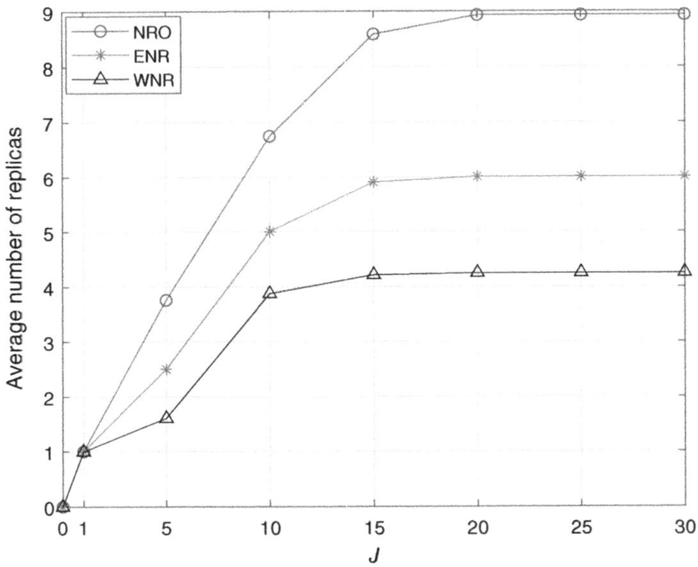

Figure 7.5 Hit ratio performance versus number of FBSs J. *Source:* [1] / IEEE / CC BY 4.0.

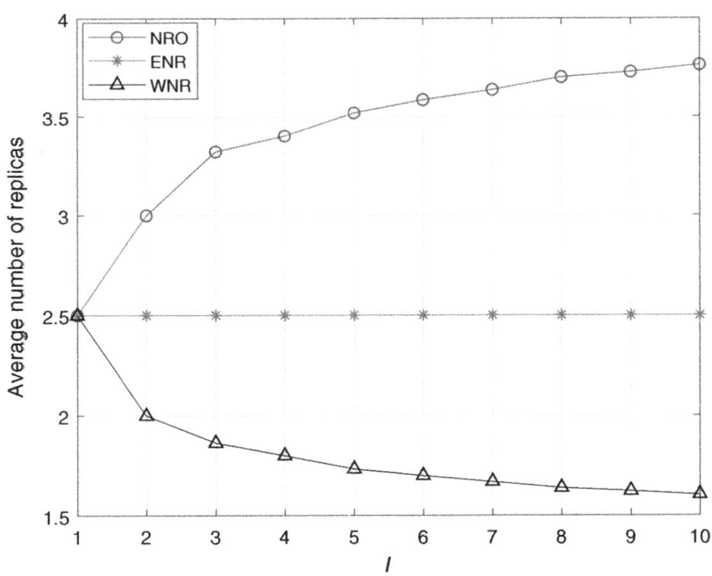

Figure 7.6 Hit ratio performance versus number of contents I. *Source:* [1] / IEEE / CC BY 4.0.

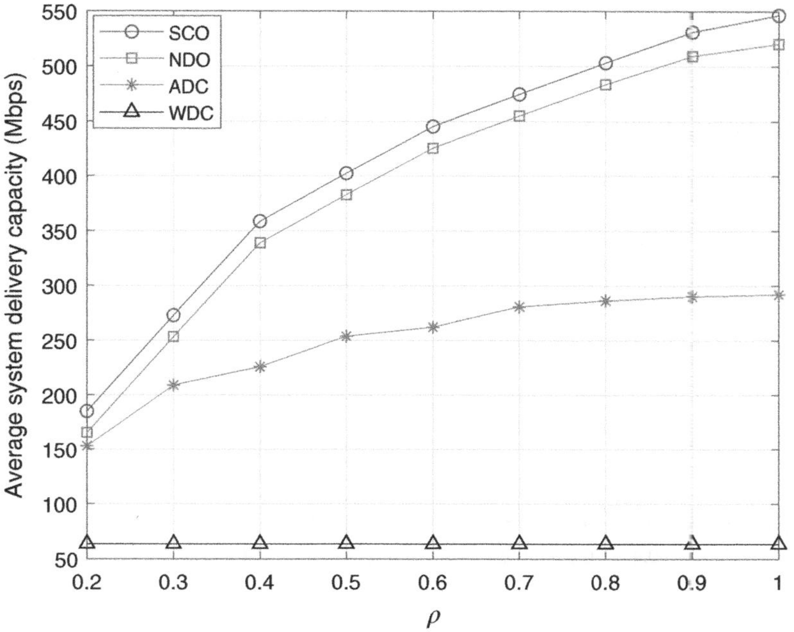

Figure 7.7 Capacity performance versus FBSs' caching capacity coefficient ρ. *Source:* [1] / IEEE / CC BY 4.0.

Figure 7.8 shows the system performance versus α. The system delivery capacity is very low in the case of NFC, indicating that femto-cell caching plays an important role in improving the system performance compared to caching in VHs and downlink resource sharing for D2D offloading. The average system delivery capacity of NFC keeps unchanged because α mostly impacts femto-cell caching rather than D2D offloading. To illustrate the impact of femto-cell caching on system performance, Figure 7.9 depicts the system delivery capacity of SCO, ADC, WDC, and NDO versus J. As can be seen, the larger number of FBSs provides higher system delivery capacity, except for the WDC. In all cases, the proposed SCO always gains the highest performance.

The impact of the number of SHs K willing to share the downlink resources is also investigated. In Figure 7.10, the system performance is higher with the increase of K for more possibilities of D2D offloading. When $K = 0$, the SCO and NDO have the same performance, meaning that there is no D2D offloading if none of the SHs shares the downlink resource. Similarly, the results of WDC and NFC are the same because there is no femto-cell caching nor D2D utilization in the NFC.

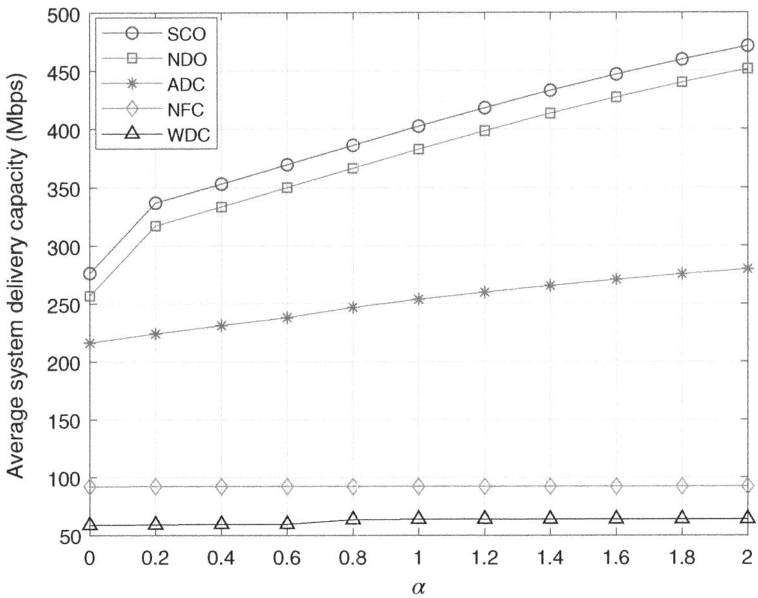

Figure 7.8 Capacity performance versus skewed access rate coefficient α. *Source:* [1] / IEEE / CC BY 4.0.

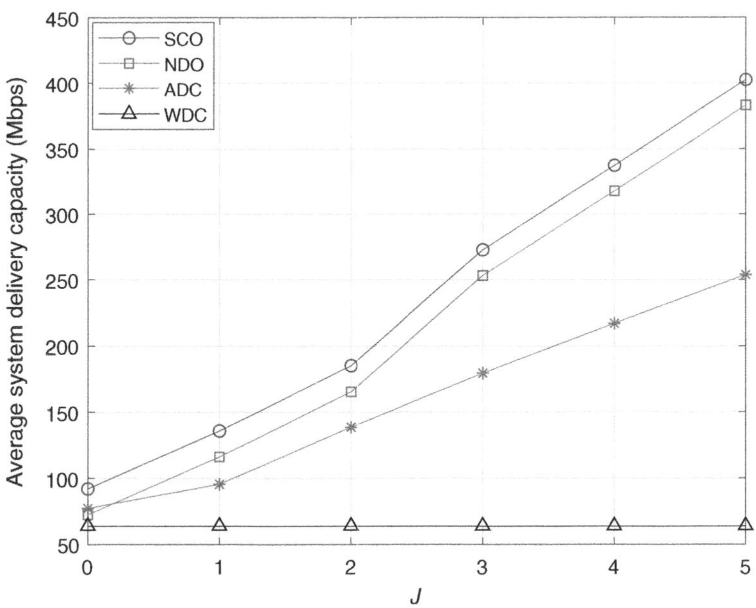

Figure 7.9 Capacity performance versus number of FBSs J. *Source:* [1] / IEEE / CC BY 4.0.

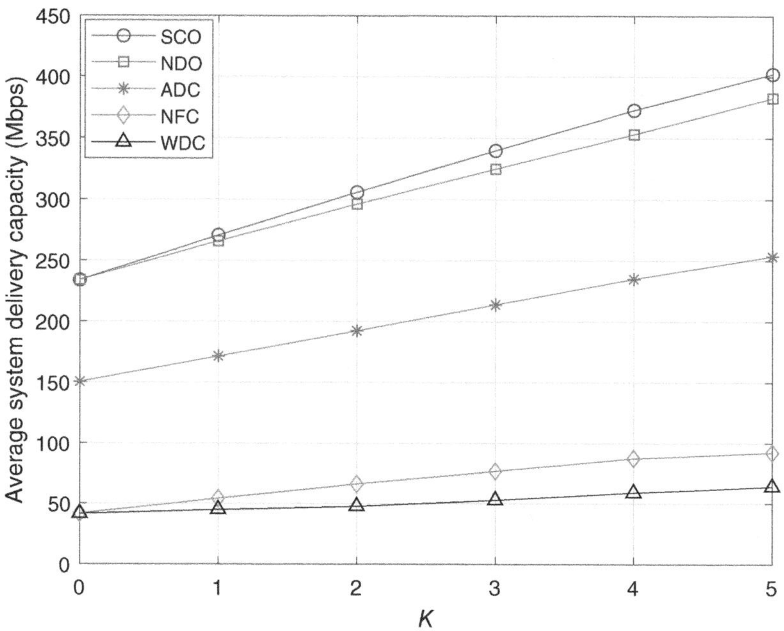

Figure 7.10 Capacity performance versus number of SHs K. *Source:* [1] / IEEE / CC BY 4.0.

Finally, we evaluate the system performance under the impact of the target SINR of the SHs by changing γ_0 from 0 dB to 30 dB. In Figure 7.11, when increasing γ_0, in the case of SCO and NFC, more D2D communications in close proximity are not exploited because they do not satisfy the target SINR of the SHs given in (7.22). The increase of γ also removes more infeasible matrices from the search spaces $\mathcal{U}$ and $\mathcal{V}$ that deteriorate the system performance. For example, when the distance from VHs to RUs for D2D communications is longer than that from VHs to SHs, it generates high interference to the SHs while not significantly improving the performance of D2D offloading. Therefore, the system performance of WDC gets better when γ increases. The performance balance between SCO and WDC keeps the performance of ADC unchanged versus γ. In addition, the system delivery capacity of NDO does not change because caching in VHs and downlink resource sharing are not considered. The results of SCO and the NFC, respectively, converge on that of NDO and WDC when γ increases.

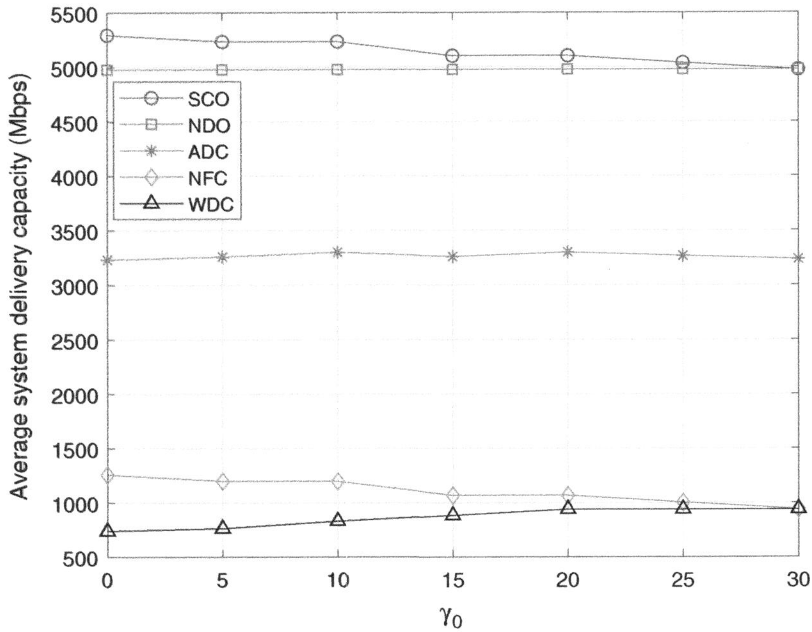

Figure 7.11 Capacity performance versus the target SINR of CUs γ.
Source: [1] / IEEE / CC BY 4.0.

7.6 Conclusion

We have designed the CDN based on the joint downlink spectrum SCF for MAS in 6G networks with integrated M-IoT. The SCF was built on exploiting the caching storage of all MBS, FBSs, and VHs, and the downlink spectrum resource of the SHs to assist the CDN. The CSF consisted of two optimization solutions: NRO and SCO, for high hit ratio and high system delivery capacity. The simulation results showed that the proposed SCF can serve the MUs the best system performance compared to other schemes.

References

1 N.-S. Vo, T. Q. Duong, M. Guizani, and A. Kortun. 5G optimized caching and downlink resource sharing for smart cities. *IEEE Access*, 6:31457–31468. 2018.

2 Y. Fang, Y. Deng, and X. Chen. Resources on the move for smart city: A disruptive perspective on the grand convergence of sensing, communications, computing, storage, and intelligence. *IEEE Commun. Mag.*, 63(4):200–206. 2025.

3 X. Chen, Y. Deng, H. Ding, G. Qu, H. Zhang, and P. Li. Vehicle as a service (VaaS): Leverage vehicles to build service networks and capabilities for smart cities. *IEEE Commun. Surv. Tuts.*, 26(3):2048–2081. 2024.

4 M. Sheraz, T. C. Chuah, Y. L. Lee, M. M. Alam, A. Al-Habashna, and Z. Han. A comprehensive survey on revolutionizing connectivity through artificial intelligence-enabled digital twin network in 6G. *IEEE Access*, 12:49184–49215. 2024.

5 N.-S. Vo, D.-B. Ha, B. Canberk, and J. Zhang. Green two-tiered wireless multimedia sensor systems: An energy, bandwidth, and quality optimisation framework. *IET Commun.*, 10(18):2543–2550. 2016.

6 N.-S. Vo, T.-H. Nguyen, and H. K. Nguyen. Joint active duty scheduling and encoding rate allocation optimized performance of wireless multimedia sensor networks in smart cities. *Springer Mobile Netw. Appl.*, 23(6):1586–1596. 2018.

7 C. M. Park, R. A. Rehman, and B.-S. Kim. Packet flooding mitigation in CCN-based wireless multimedia sensor networks for smart cities. *IEEE Access*, 5:11054–11062. 2017.

8 S. P. Singh, N. Kumar, N. S. Alghamdi, G. Dhiman, W. Viriyasitavat, and A. Sapsomboon. Next-gen WSN enabled IoT for consumer electronics in smart city: Elevating quality of service through reinforcement learning-enhanced multi-objective strategies. *IEEE Trans. Consum. Electron.*, 70(4):6507–6518. 2024.

9 T.-V. Nguyen, N.-N. Dao, V. D. Tuong, W. Noh, and S. Cho. User-aware and flexible proactive caching using LSTM and ensemble learning in IoT-MEC networks. *IEEE Internet Things J.*, 9(5):3251–3269. 2022.

10 A. Boukerche, R. W. L. Coutinho, and A. A. F. Loureiro. Information-centric cognitive radio networks for content distribution in smart cities. *IEEE Netw.*, 33(3):146–151. 2019.

11 H. Li, X. Li, C. Sun, F. Fang, Q. Fan, and X. Wang. Intelligent content caching and user association in mobile edge computing networks for smart cities. *IEEE Trans. Netw. Sci. Eng.*, 11(1):994–1007. 2024.

12 N.-S. Vo, T. Q. Duong, and M. Guizani. Quality of sustainability optimization design for mobile ad hoc networks in disaster areas. In *Proc. IEEE Global Commun. Conf.*, pages 1–6, San Diego, CA, Dec. 2015.

13 W. Jiang and F.-L. Luo. Cellular and Cell-Free Massive MIMO Techniques in 6G. In *6G Key Technologies: A Comprehensive Guide*, Wiley-IEEE Press, 2023.

14 D. G. Riviello, R. Tuninato, and R. Garello. Multi-layer multi-user MIMO with cylindrical arrays under 3GPP 3D channel model for B5G/6G networks. *IEEE Access*, 12:145753–145767. 2024.

15 L. Breslau, P. Cao, L. Fan, G. Phillips, and S. Shenker. Web caching and Zipf-like distributions: Evidence and implications. In *Proc. IEEE Int. Conf. on Computer Commun.*, pages 126–134, New York, NY, Mar. 1999.

16 V. Chandrasekhar and J. G. Andrews. Spectrum allocation in tiered cellular networks. *IEEE Trans. Commun.*, 57(10):3059–3068. 2009.

17 W. C. Cheung, T. Q. S. Quek, and M. Kountouris. Throughput optimization, spectrum allocation, and access control in two-tier femtocell networks. *IEEE J. Sel. Areas Commun.*, 30(3):561–574. 2012.

18 S. Mehrotra. On the implementation of a primal-dual interior point method. *SIAM J. Optim.*, 2:575–601. 1992.

19 Y. Zhang. Solving large-scale linear programs by interior-point methods under the MATLAB environment. *Optim. Method. Softw.*, 10(1):1–31. 2007

20 N.-S. Vo, T. Q. Duong, and M. Guizani. QoE-oriented resource efficiency for 5G two-tier cellular networks: A femtocaching framework. In *Proc. IEEE Global Commun. Conf.*, pages 1–6, Washington, DC, Dec. 2016.

21 X. Cai, J. Zheng, Y. Zhang, and H. Murata. A capacity oriented resource allocation algorithm for device-to-device communication in mobile cellular networks. In *Proc. IEEE Inter. Commun. Conf.*, pages 2233–2238, Sydney, Australia, Jun. 2014.

8

Joint Energy, Bandwidth, and Quality Optimization for Multimedia Applications and Services in Content Delivery Networks

8.1 Introduction

Together with other industries, the development of wireless communications has introduced a rapid increase in energy consumption that causes many existing and potential problems.[1] In the era of 6G, the high number of MUs demanding ubiquitous multimedia applications and services (MAS) and the proliferation of base stations (BSs) are considerably consuming a large portion of the energy of network infrastructure and mobile devices. Generally, wireless data traffic increases tenfold every five years, yielding a 20% increase in energy consumption, of which 79% is mobile multimedia traffic [2], and thus making up to 16% increase in energy consumption.

For the Internet service and content providers, it will be more challenging to serve the MUs with a high quality of experience (QoE) once the number of mobile-connected devices exceeds the world's population. It is evident that the industry urgently requires a sustainable multimedia communication solution to bridge the gap between quantity (i.e., number of users/devices and BSs), quality (i.e., QoE), and energy consumption. The shortage of radio frequency spectrum and bandwidth, the rising energy consumption, and the proliferation of mobile-connected devices and BSs have emerged as critical challenges worldwide. Particularly, achieving tenfold higher energy efficiency than 5G for green communications remains a major challenge in 6G networks [3].

In smart cities of information and communication technology theme, the number of MUs enjoying MAS increases quickly. This requires the CDNs to be more powerful and flexible so that the MUs can have multiple selections of different channel accesses, edge caching networks, and edge servers to obtain the best MAS. To do so, the conventional architecture of CDN has to be modified by expanding closer to the MUs in the form of edge networks [4, 5]. While the convergence

1 This chapter has been published partly in [1].

of digital twins, Internet of Things (IoT), artificial intelligence, and 6G networks certainly addresses the urgent environmental crisis well, an edge network can be optimally designed to assist conventional CDN in progressing toward the global agreement of zero carbon emissions [6–8].

In this chapter, we propose a simple M-IoT system-assisted CDN in 6G networks. The CSs capture the videos, then disseminate them to the edge network, i.e., the CHs and BSs, for caching. The edge caching network (ECN) can help the CDN provide the MUs with cached videos. In other words, the M-IoT system can collect the videos of ambient phenomena and cache them at the ECN for future use. The system consists of an EBQ framework that considers many aspects of videos (size, popularity, and RDM), resources of CSs and CHs (energy and bandwidth), lossy features of wireless channels, and playback quality of videos requested by the MUs. The EBQ includes an ARA optimization problem formulated and solved by genetic algorithms (GAs) to find the optimal encoding rates for each captured video. Based on the optimal encoding rates, the videos are then packetized by applying the LMDC-FEC for dissemination.

It is important to note that the LMDC-FEC [9–12] is applied to scalable extensions of high efficiency video coding (SHVC) [13]. This way, the LMDC-FEC-based SHVC allows us to packetize the videos into multiple descriptions for transmission in order to achieve the highest data protection against error and delay in multi-hop communications from the CSs to the ECN and to adapt to the diverse limited bandwidth of the M-IoT system. The objective is to minimize the energy consumption for disseminating the videos from the CSs to the ECN while utilizing the bandwidth resources at the CHs and ensuring the playback quality of the cached videos. Consequently, simulation results are analyzed and discussed to demonstrate the benefits of the proposed EBQ framework compared to other benchmarks.

The rest of this chapter is organized as follows. We introduce the M-IoT system model with the EBQ framework in Section 8.2. In Section 8.3, the system is then analyzed and formulated in detail. Section 8.4 presents the ARA optimization problem and solution using genetic algorithms (GAs). In Section 8.5, simulation results are shown to evaluate the performance of the proposed EBQ framework. Finally, we conclude the chapter in Section 8.6.

8.2 M-IoT System with EBQ Framework

We consider a typical architecture of M-IoT system as shown in Figure 8.1. Table 8.1 summarizes the key notations used in the system. The system consists of one BS and C CHs. Each CH i, which handles S_i CSs (denoted by $CS_{i,j}$), $i = 1, 2, ..., C$, $j = 1, 2, ..., S_i$, is in charge of collecting the video contents from its catastrophe-prone area and then forwarding them to the BS for caching and

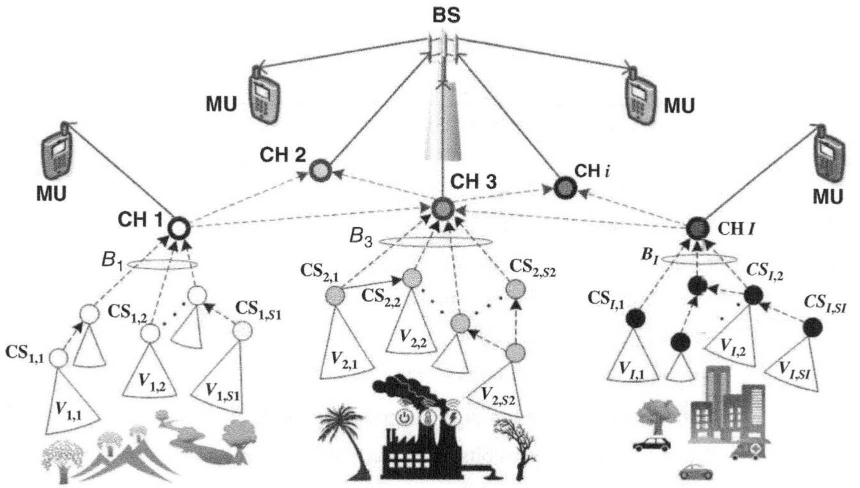

Figure 8.1 Architecture of M-IoT system.

monitoring. In this way, the M-IoT system is arranged in two tiers. The 1st tier is comprised of the static or mobile CSs with lower available energy and processing capacity. The CSs carry out simple tasks of collecting video contents from the environment, analyzing the RDM, and encoding the video contents into descriptions for sending to the BS via the 2nd tier. The 2nd tier includes the CHs, which are in charge of performing complicated tasks, e.g., controlling the CSs in the 1st tier, solving the ARA of EBQ, and forwarding the descriptions of captured videos to the BS. We assume that the CHs with higher available energy and processing capacity are elected by using a lightweight and dependable trust system for clustered wireless sensor networks given in [14]. Multi-hop interconnections among CHs and among CSs are established to ensure high connectivity in the M-IoT system.

The EBQ designed for the M-IoT system is illustrated in Figure 8.2 with main objective of formulating and solving the ARA optimization problem to minimize the average energy consumption per CS for long lifetime of the M-IoT system while conserving backhaul bandwidth resource of the CHs and guaranteeing high quality of received videos at the BS. To provide the videos to the MUs, the BS frequently performs the EBQ in three steps as follows:

- Step 1: The BS informs $CS_{i,j}$ to capture $V_{i,j}$ at full resolution $R_{i,j}$ depending on its integrated camera hardware configuration, in a predefined duration. All the captured videos are analyzed to obtain their RDMs by the CSs. Then, $CS_{i,j}$ sends the information of the RDM of $V_{i,j}$ to the BS via CH i.

Table 8.1 Notations.

Symbols	Descriptions
C	Number of cluster heads (CHs)
S_i	Number of sensor cameras (CSs) of CH i, $i = 1, 2, ..., C$
$CS_{i,j}$	CS j of CH i, $j = 1, 2, ..., S_i$
$V_{i,j}$	Video j captured by $CS_{i,j}$
M	Number of layers/descriptions of each video
$R_{i,j}$	Upper bound of encoding rate, i.e., full resolution of $V_{i,j}$
$L_{i,j}$	Size of $V_{i,j}$ measured in Kbit at full resolution
$R_{i,j}^{(m)}$	Optimal encoding rate of $V_{i,j}$, $m = 1, 2, ..., M$
$D_{i,j}(R_{i,j}^{(m)})$	Reconstructed distortion of $V_{i,j}$ encoded at rate $R_{i,j}^{(m)}$
$L_{i,j}^{FEC}$	Size of $V_{i,j}$ measured in bits after encoding with embedded forward error correction (FEC)
$\overline{E}_{i,j}$	Average energy consumption per CS to transmit $V_{i,j}$ from $CS_{i,j}$ to CH i
$\overline{B}_i$	Average bandwidth served by CH i
$H_{i,j}^{(1)}$	Number of hops to transmit $V_{i,j}$ from the $CS_{i,j}$ to CH i in the 1st tier
$H_i^{(2)}$	Number of hops to forward $V_{i,j}$ from CH i to the BS in the 2nd tier
$e_{i,j}^{(1,h)}$	Description error rate of the h-th hop on the route from $CS_{i,j}$ to CH i in the 1st tier
$e_i^{(2,k)}$	Description error rate of the k-th hop on the route from CH i to the BS in the 1st tier
$p_{i,j}^{(m)}$	Probability that there are m out of M description of $V_{i,j}$ correctly received

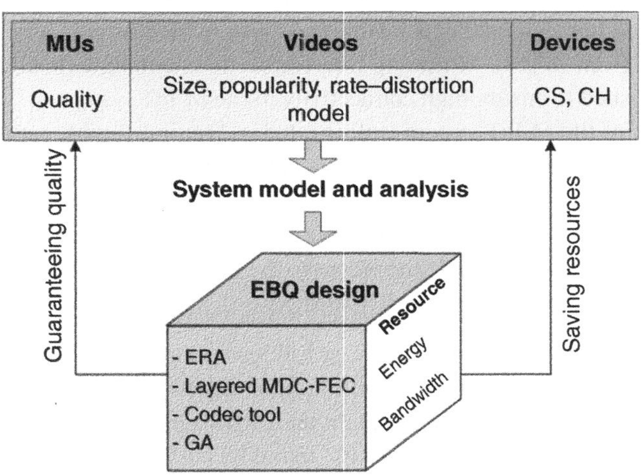

Figure 8.2 EBQ optimization framework.

- Step 2: The BS collects energy conditions of the CSs, backhaul bandwidth guarantees of the CHs, and channel state information (CSI), e.g., lossy features of wireless channels, from the whole system. Together with the RDMs, the BS formulates the ARA and solves it for the optimal encoding rates $R_{i,j}^{(m)}$, allocated to $V_{i,j}$, here $m = 1, 2, ..., M$ and $R_{i,j}^{(M)} \leq R_{i,j}$. The optimal rate results $R_{i,j}^{(m)}$ are sent back to the corresponding $CS_{i,j}$ for packetizing.
- Step 3: Based on the optimal rate results, $CS_{i,j}$ packetizes $V_{i,j}$ into multiple descriptions for transmission to the BS via CH i.

8.3 Analysis and Formulation of EBQ Optimization Framework

8.3.1 Source Video Packetization Scheme

In our EBQ, the ARA is formulated based on the RDMs of the captured videos sent from the CSs. We apply the RDM introduced in [15] to SHVC [13]. In this RDM, the source reconstructed distortion $D_{i,j}(R_{i,j}^{(m)})$ of $V_{i,j}$ is a decaying exponential function of encoding rate (Kbps), given by

$$D_{i,j}(R_{i,j}^{(m)}) = \gamma_{i,j}(R_{i,j}^{(m)})^{\beta_{i,j}}, \tag{8.1}$$

where $\gamma_{i,j}$ and $\beta_{i,j}$ are the dependent parameters of $V_{i,j}$. In step 1 of EBQ, the captured videos are analyzed to obtain their RDMs featured by $\gamma_{i,j}$ and $\beta_{i,j}$.

Recently, SHVC has been developed to exploit the benefits of scalable H.264/advanced video coding (AVC) [13]. We utilize SHVC to cope with the diverse resolution of CSs and the condition of wireless channels in the two-tiered M-IoT system. To further overcome the problems of lossy transmission and multi-hop delay, we apply LMDC-FEC [9–12] to SHVC to packetize the captured videos into descriptions for transmission. Although LMDC-FEC-based SHVC yields higher coding complexity than pure scalable AVC, it meets the current road map of processing technology, which has been developed faster than other technologies. The LMDC-FEC-based SHVC packetization scheme is presented below.

Consider $V_{i,j}$ and its RDM as shown in Figure 8.3, after obtaining the optimal rates $R_{i,j}^{(m)}$ from step 2 of EBQ such that $0 = R_{i,j}^{(0)} \leq R_{i,j}^{(1)} \leq R_{i,j}^{(2)} \leq ... \leq R_{i,j}^{(m)} \leq ... \leq R_{i,j}^{(M-1)} \leq R_{i,j}^{(M)} \leq R_{i,j}$, $V_{i,j}$ is divided into M layers [12]. Layer m is denoted by two consecutive rate points $R_{i,j}^{(m-1)}$ and $R_{i,j}^{(m)}$. Thus, the reconstructed distortion of $V_{i,j}$ depends not only on the number of correctly received layers but also on their priority order. If layer m is not correctly received, all layers from $m + 1$ to M become unusable even if they are correctly received.

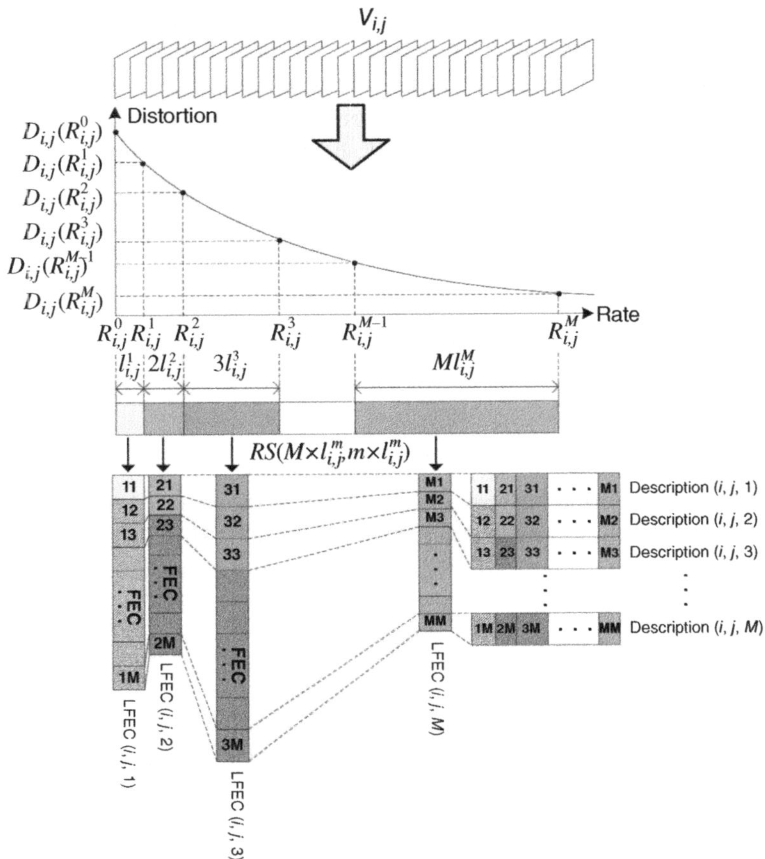

Figure 8.3 LMDC-FEC-based SHVC packetization scheme. *Source:* Reproduced from [1] / John Wiley & Sons.

To remove the priority order for higher performance of reconstruction, layer m is then partitioned into m blocks, each with the same size of

$$l_{i,j}^{(m)} = \left\lfloor \frac{(R_{i,j}^{(m)} - R_{i,j}^{(m-1)})L_{i,j}}{mR_{i,j}} \right\rfloor, \tag{8.2}$$

where $L_{i,j}$ is the original size of $V_{i,j}$ measured in bits and the floor operator $\lfloor . \rfloor$ is used to round a real number to the nearest integer less than or equal to it. This operator may result in $\sum_{m=1}^{M} ml_{i,j}^{(m)} < L_{i,j}$, but it does not seriously affect the performance of the packetization scheme because the total reduction in bits is infinitesimal compared to $L_{i,j}$. For simplicity, we do not focus on the bit rate adjustment algorithm to compensate for bits dropped by the floor operator.

Next, a Reed–Solomon erasure code $RS(M \times l_{i,j}^{(m)}, m \times l_{i,j}^{(m)})$ is applied to m original blocks (of size $m \times l_{i,j}^{(m)}$ bits) of layer m to generate the codewords of length M blocks (of size $M \times l_{i,j}^{(m)}$ bits) with embedded FEC. The $RS(M \times l_{i,j}^{(m)}, m \times l_{i,j}^{(m)})$ is able to correct up to $M - m$ erroneous blocks of layer m of $V_{i,j}$. In other words, the first m layers can be reconstructed if m out of M blocks are correctly received.

Finally, to generate M descriptions of $V_{i,j}$ for transmission, the m-th blocks from all layers of $V_{i,j}$ are selected and packetized into description m. To this end, if m out of M descriptions of $V_{i,j}$ are correctly received (regardless of their priority order), the first m layers of $V_{i,j}$ are reconstructed and thus the $V_{i,j}$ is played back at rate $R_{i,j}^{(m)}$ with corresponding reconstructed distortion $D_{i,j}(R_{i,j}^{(m)})$. The total size of $V_{i,j}$ with embedded FEC is given by

$$L_{i,j}^{\mathrm{FEC}} = M \sum_{m=1}^{M} l_{i,j}^{(m)}. \tag{8.3}$$

8.3.2 Energy Consumption

The main objective of ARA in EBQ is to minimize energy consumption in order to prolong the lifetime of the M-IoT system. To this end, we assume that the available energy levels of CHs are significantly higher than those of CSs. Consequently, we focus only on the available energy levels of CSs to compute the average energy consumption per CS, which serves as the objective function to be minimized in ARA.

Let $H_{i,j}^{(1)}$ be the number of hops on the path from $CS_{i,j}$ to CH i and $d_{i,j}^{(h)}$ be the wireless distance measured in meters of hop h. Following [16], the energy consumption to transmit a bit over $d_{i,j}^{(h)}$ meters is given by

$$E_{i,j}^{(h)} = \theta_1 + \theta_2 (d_{i,j}^{(h)})^{\eta}, \tag{8.4}$$

where $\theta_1 = 50$nJ/bit/hop, $\theta_2 = 0.0013$ pJ/bit/m^4/hop, and $\eta = 4$.

To transmit $L_{i,j}^{\mathrm{FEC}}$ bits of $V_{i,j}$ over $H_{i,j}^{(1)}$ hops, the average energy consumption per CS from $CS_{i,j}$ to CH i is expressed as

$$\overline{E}_{i,j} = \frac{L_{i,j}^{\mathrm{FEC}}}{H_{i,j}^{(1)}} \sum_{h=1}^{H_{i,j}^{(1)}} E_{i,j}^{(h)}. \tag{8.5}$$

In (8.5), we only consider the energy drained from transmission, not from processing. Intermediate CSs on the path from $CS_{i,j}$ to CH i are established by applying utility energy-based opportunistic routing protocol for lifetime enhancement [17]. We further assume that the lifetime of this path is much greater than the streaming session of $V_{i,j}$. The overall average energy consumption for transmission of all videos per CS can be written as

$$\overline{E} = \frac{1}{C}\sum_{i=1}^{C}\sum_{j=1}^{S_i} r_{i,j}\overline{E}_{i,j}, \tag{8.6}$$

where $r_{i,j}$ reflects the priority/importance of $V_{i,j}$ captured from the catastrophe-prone area by $\text{CS}_{i,j}$. The value of $r_{i,j}$ follows Zipf-like distribution [18] as

$$r_{i,j} = \frac{j^{-\alpha_i}}{\sum_{j'=1}^{S_i} (j')^{-\alpha_i}}, \tag{8.7}$$

where $\alpha_i \geq 0$ represents the skewed importance between different videos handled by CH i and $\sum_{j=1}^{S_i} r_{i,j} = 1$. If $\alpha_i >> 0$, the $V_{i,j}$ is much more important than the $V_{i,j+1}$, otherwise, if $\alpha_i = 0$, all videos of CH i have the same importance. Zipf-like distribution is used to ensure that in each CH, the video contents with higher importance are transmitted with less energy consumption while conserving bandwidth resource and guaranteeing high quality.

8.3.3 Bandwidth Consumption

The problem of bottleneck at the CHs in the M-IoT system may occur if we do not carefully consider the maximum encoding rate $R_{i,j}^{(M)}$ of $V_{i,j}$ during the encoding process. The solution is that we have to derive the average bandwidth coming from the $\text{CS}_{i,j}$ to CH i and then limit this average value by keeping it within a given backhaul bandwidth of CH i in the bandwidth constraint of ARA. The average bandwidth served by CH i can be given by

$$\overline{B}_i = \sum_{j=1}^{S_i} r_{i,j}R_{i,j}^{(M)}. \tag{8.8}$$

8.3.4 Quality Guarantee

For high quality guarantee, we have to calculate the average reconstructed distortion of received videos at the BS, which is considered as the quality constraint of ARA. Let $e_{i,j}^{(1,h)}$ be the description error rate of hop h on the path from $\text{CS}_{i,j}$ to CH i in the 1st tier and $e_i^{(2,k)}$ be the description error rate of hop k on the path from

CH i to the BS in the 2nd tier, the description error probability over the path from the $\text{CS}_{i,j}$ to the BS is expressed as

$$e_{i,j} = 1 - \left[\prod_{h=1}^{H_{i,j}^{(1)}} (1 - e_{i,j}^{(1,h)}) \prod_{k=1}^{H_i^{(2)}} (1 - e_i^{(2,k)}) \right], \tag{8.9}$$

where $H_i^{(2)}$ is the number of hops on the path from CH i to the BS.

As analyzed, if there are m out of M descriptions correctly received, $V_{i,j}$ is reconstructed and played back at rate $R_{i,j}^{(m)}$ with distortion $D_{i,j}(R_{i,j}^{(m)})$. Let $p_{i,j}^{(m)}$ be the probability that m out of M descriptions of $V_{i,j}$ are correctly received, the reconstructed distortion of $V_{i,j}$ can be given by

$$D_{i,j} = \sum_{m=0}^{M} p_{i,j}^{(m)} D_{i,j}(R_{i,j}^{(m)}), \tag{8.10}$$

where

$$p_{i,j}^{(m)} = \binom{m}{M} (1 - e_{i,j})^m (e_{i,j})^{M-m}. \tag{8.11}$$

Taking the importance value $r_{i,j}$ of $V_{i,j}$ captured by $\text{CS}_{i,j}$ into account, the overall average reconstructed distortion of all videos received at the BS is

$$\overline{D} = \frac{1}{C} \sum_{i=1}^{C} \sum_{j=1}^{S_i} r_{i,j} D_{i,j}. \tag{8.12}$$

8.4 ARA Optimization Problem and Solution

8.4.1 ARA Optimization Problem

In ARA, the average energy consumption (8.6) is considered as the objective function, which is minimized to prolong the lifetime of the CSs in the M-IoT system by finding the optimal rates $R_{i,j}^{(m)}$ allocated to each video. In addition, to conserve the backhaul bandwidth of the CHs and guarantee high quality of received videos at the BS, the average bandwidth $\overline{B}_i$ (8.8) served by CH i and the average reconstructed distortion $\overline{D}$ (8.12) at the BS, which cannot be greater than given thresholds, are taken into account in the constraints of ARA. Mathematically, ARA is formulated as follows:

$$\min_{R_{i,j}^{(m)}} \overline{E}, \tag{8.13}$$

$$s.t.\begin{cases} R_{i,j}^{(0)} \leq R_{i,j}^{(1)} \leq R_{i,j}^{(2)} \leq \dots \leq R_{i,j}^{(m)} \leq \dots \leq R_{i,j}^{(M-1)} \leq R_{i,j}^{(M)}, \\ R_{i,j}^{(M)} \leq R_{i,j}, \\ \overline{B}_i \leq \delta_B B_i^*, \\ \overline{D} \leq \delta_D D^*. \end{cases} \tag{8.14}$$

In (8.14), the first two constraints are to satisfy the rate allocation characteristic discussed in Section 8.3.1. B_i^* and D^* in the last two constraints, which are the backhaul bandwidth of CH i and the maximum average reconstructed distortion, are used together with two adjustable coefficients $\delta_B > 0$ (bandwidth consumption coefficient) and $\delta_D > 0$ (quality guarantee coefficient) to limit the average bandwidth served by the CH i and guarantee high quality of received videos at the BS, respectively.

Though (8.13) and (8.14) can be converted to an unconstrained optimization problem and then solved by applying Lagrange multipliers [19], the complexity of Lagrangian-based solution for global optimal results of ARA is prohibitively high [12, 15]. We therefore introduce GA to solve ARA in the sequel.

8.4.2 Solution with GA

In practical systems, deploying a global and precise optimization solution with excessive time and computational complexity is not feasible compared to the one within reasonable complexity. For exactly or approximately optimal solutions within a reasonable time frame, a class of adaptive heuristic algorithms, i.e., GA [20], based on the natural evolutionary principles of selection, variation, and inheritance, can be applied as a promising method. However, the problem is that the constraints in (8.14) are more complicated than the ones in the form of simple lower and upper bounds. This may cause GA to result in infeasible offsprings that do not satisfy the constraints. To overcome this, penalty function approach is used to remove the constraints in (8.13) and (8.14) [21]. In this way, the constraints in (8.14) are rewritten in sequence as

$$\begin{cases} \Delta R_{i,j}^{(l)} = R_{i,j}^{(l+1)} - R_{i,j}^{(l)} \geq 0, l = 0, 1, 2, \dots, M-1, \\ \Delta R_{i,j} = R_{i,j} - R_{i,j}^{(M)} \geq 0, \\ \Delta \overline{B}_i = \delta_B B_i^* - \overline{B}_i \geq 0, \\ \Delta \overline{D} = \delta_D D^* - \overline{D} \geq 0. \end{cases} \tag{8.15}$$

Then, we obtain the penalty function given by

$$F_\mathrm{P} = \lambda_1 \sum_{i=1}^{C} \sum_{j=1}^{S_i} \sum_{l=0}^{M-1} [\min\{0, \Delta R_{i,j}^{(l)}\}]^2 + \lambda_2 \sum_{i=1}^{C} \sum_{j=1}^{S_i} [\min\{0, \Delta R_{i,j}\}]^2$$
$$+\lambda_3 \sum_{i=1}^{C} [\min\{0, \Delta \overline{B}_i\}]^2 + \lambda_4 [\min\{0, \Delta \overline{D}\}]^2, \tag{8.16}$$

where $\lambda_1, \lambda_2, \lambda_3$, and λ_4 are added to reflect the violation degree of the constraints. Pulling (8.13) and (8.16) together, we get the following unconstrained optimization problem

$$\min_{R_{i,j}^{(m)}} \overline{E}_\mathrm{P} = \overline{E} + F_\mathrm{P}. \tag{8.17}$$

GA can be applied to solving (8.17) instead of (8.13) and (8.14). The detailed implementation of GA for the ARA is presented in Algorithm 8.1.

Algorithm 8.1 Implementation of GA

Input: Initial parameters of GA:
- $N_\mathrm{P} = 500$: Population size, i.e., number of individuals
- $N_\mathrm{B} = 50$: Number of bits to represent an individual in the solution set $\{C_z\} = \{R_{i,j}^{(m,z)}\}$, $z = 1, 2, \dots, N_\mathrm{P}$
- $N_\mathrm{G} = 50$: Number of generations
- $P_\mathrm{G} = 0.9$: Generation gap
- $P_\mathrm{C} = 0.1$: Crossover probability
- $P_\mathrm{M} = 10^{-6}$: Mutation probability

Output: $C^* = R_{i,j}^{(m)*}$

Generate N_P random individuals of solution $\{C_z\}$
Calculate fitness values $\overline{E}_\mathrm{P}(\{C_z\})$ in (8.17) corresponding to each individual in $\{C_z\}$ of current generation
while $N_\mathrm{G} > 0$ **do**
    Put $\{C_z\}$ and $\overline{E}_\mathrm{P}(\{C_z\})$ in mating pool
    Select $N_\mathrm{P} \times P_\mathrm{G}$ best individuals with better fitness values, i.e., lower values of $\overline{E}_\mathrm{P}(\{C_z\})$, for breeding the next generation by applying stochastic universal sampling operator
    Collect 2 parent individuals to generate 2 offsprings by applying single point crossover operator with crossover probability P_C
    Mutate 2 new offsprings with mutation probability P_M to recover the good genes that could be lost due to the operators in previous steps
    Calculate the fitness values of all offsprings and reinserting them into the current population
    $N_\mathrm{G} \leftarrow N_\mathrm{G} - 1$
end while
Find the best fitness and related individual $\overline{E}_\mathrm{P}(C^*)$ in the last generation

---

Table 8.2 Parameter settings.

Notations	Specifications
C	3 CHs
S_i	3 CSs or 3 video contents captured in each cluster
M	50 layers/descriptions
$H_{i,j}^{(1)}, H_i^{(2)}$	Randomly distributed from 1 to 5 hops
$d_{i,j}^{(h)}$	Randomly distributed from 10 to 50 meters
$e_{i,j}^{(1,h)}, e_i^{(2,k)}$	Randomly distributed from 0.0005 to 0.05
$\alpha\{\alpha_i\}$	$\alpha\{1, 2, 3\}$, α changes from 0 to 1 to adjust α_i
$\{B_i^*\}$	$\{10,000,\ 100,000,\ 30,000\}$ Kbps, obtained by approximately averaging the full resolutions $R_{i,j}$ in each CH as shown in Table 8.3
D^*	1000 (MSE)
δ_B	From 1 to 10 to adjust $\{B_i^*\}$
δ_D	From 0.1 to 1 to adjust D^*

Table 8.3 Information of videos.

$V_{i,j}$	Pixels|Frames	$\gamma_{i,j}\,\vert\beta_{i,j}$	$R_{i,j}$ (Kbps)	$L_{i,j}$ (Kbit)
$V_{1,1}$ Foreman	352 × 288|300	9,806e+00|-0.9972	3,000	$L_{1,1} = 35{,}600$
$V_{1,2}$ Mobile	352 × 288|300	3,652e+05|-1.9690	9,000	$L_{1,2} = 35{,}600$
$V_{1,3}$ Basketballdrill	832 × 480|501	6,566e+01|-0.9714	20,000	$L_{1,3} = 2{,}344{,}680$
$V_{2,1}$ Peopleonstreet	2560 × 1600|21	1,197e+03|-1.1160	105,000	$L_{2,1} = 1{,}015{,}808$
$V_{2,2}$ Traffic	2560 × 1600|150	1,644e+03|-1.0920	170,000	$L_{2,2} = 113{,}496$
$V_{2,3}$ Blowingbubbles	416 × 240|97	2,031e+02|-1.1320	9,600	$L_{2,3} = 7{,}200{,}000$
$V_{3,1}$ Tennis	1920 × 1080|240	3,820e+00|-0.6706	40,500	$L_{3,1} = 5{,}832{,}000$
$V_{3,2}$ Racehorses	832 × 480|97	1,108e+03|-1.2060	28,300	$L_{3,2} = 453{,}960$
$V_{3,3}$ Basketballpass	416 × 240|300	7,652e+01|-1.1530	5,300	$L_{3,3} = 351{,}000$

8.5 Performance Evaluation

8.5.1 Parameters Setting

We deploy the M-IoT system with the parameters in Tables 8.2 and 8.3. The considered videos are analyzed to generate their RDMs by using HM reference software version 12.0 [22].

8.5.2 Convergence Rate of GA

The convergence rate of GA is shown in Figure 8.4. We implement 100 generations, and convergence is achieved within about 50 generations, when the mean fitness of all individuals reaches the best fitness value. The GA produces better individuals in successive generations by mimicking natural evolution. This convergence rate demonstrates that GA is a feasible solution for the ARA.

8.5.3 Performance Metrics

We evaluate the proposed EBQ by comparing with two schemes: equal rate allocation (ERA) and random rate allocation (RRA). In ERA scheme, the rate $R_{i,j}^{(m),\mathrm{ERA}}$ is equally distributed from 0 to $R_{i,j}^{(M)}$ such that $R_{i,j}^{(m),\mathrm{ERA}} = \frac{mR_{i,j}^{(M)}}{M}$. Meanwhile, the rate $R_{i,j}^{(m),\mathrm{RRA}}$ in RRA scheme is randomly distributed from 0 to $R_{i,j}^{(M)}$ such that $0 \leq R_{i,j}^{(1),\mathrm{RRA}} \leq R_{i,j}^{(2),\mathrm{RRA}} \leq \ldots \leq R_{i,j}^{(m),\mathrm{RRA}} \leq \ldots \leq R_{i,j}^{(M-1),\mathrm{RRA}} \leq R_{i,j}^{(M),\mathrm{RRA}}$. It means that the average bandwidth consumption of EBQ, ERA, and RRA is the same as expressed in (8.8). Therefore, only energy and quality metrics of EBQ, ERA, and RRA are evaluated. To assess the quality performance, the distortion

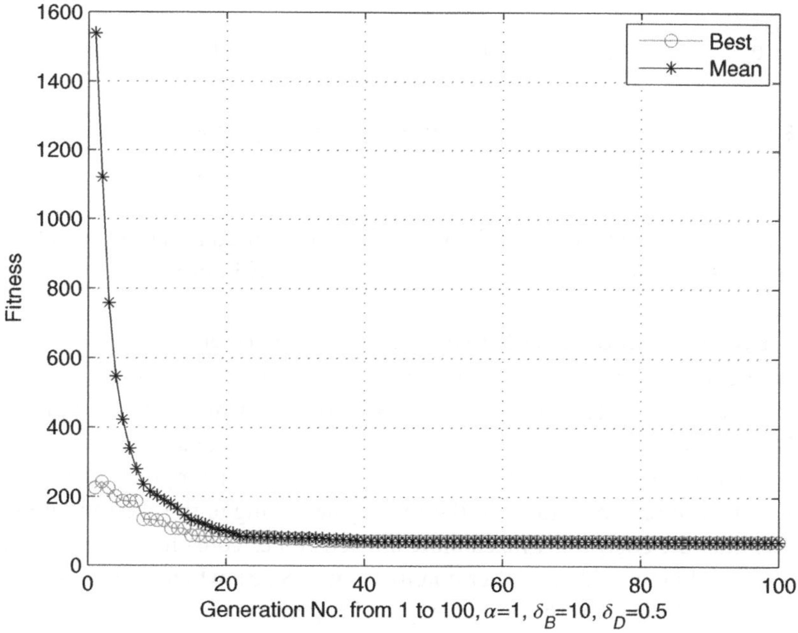

Figure 8.4 Convergence rate of GA. *Source:* Reproduced from [1] / John Wiley & Sons.

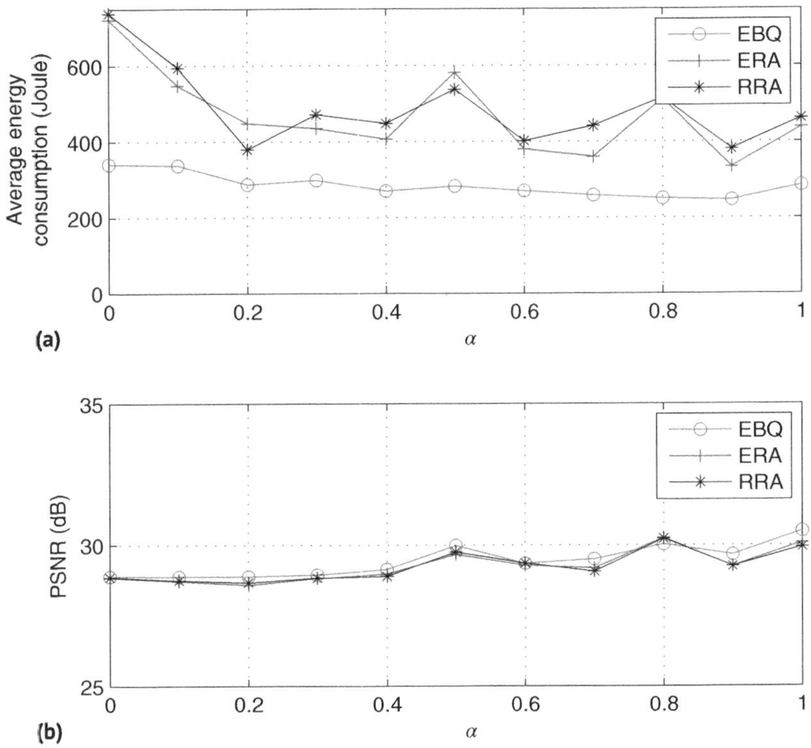

Figure 8.5 Performance evaluation versus α: (a) average energy consumption; (b) PSNR. *Source:* Reproduced from [1] / John Wiley & Sons.

initially quantified by mean squared error (MSE) is converted into peak signal-to-noise ratio (PSNR), as the latter offers a more perceptually interpretable measure.

8.5.3.1 EBQ Performance Versus Skewed Importance Parameter α_i

To illustrate the energy and quality performance metrics, we keep $\delta_B = 10$ and $\delta_D = 0.1$ and change the skewed importance of videos by multiplying $\{\alpha_i\} = \{1, 2, 3\}$ by a coefficient α, $0 \leq \alpha \leq 1$, as shown in Table 8.2. Figure 8.5 confirms that in the proposed EBQ, the average energy consumption of the CSs and the quality of received videos at the BS vary according to the skewed importance such that the energy is utilized more efficiently, i.e., consuming less energy (Figure 8.5a) while providing higher quality (Figure 8.5b), when α increases. In comparison, the EBQ gains the best energy-saving performance and mostly guarantees higher quality compared to the ERA and RRA schemes.

8.5.3.2 EBQ Performance Versus Quality Guarantee Coefficient δ_D

Next, we evaluate the performance of EBQ, ERA, and RRA by changing δ_D from 0.1 to 1 while keeping $\delta_B = 10$ and $\alpha = 1$. Figure 8.6 shows that when the quality demand decreases (i.e., increase δ_D), the CSs adaptively require less energy consumption (Figure 8.6a) to meet the decrease of quality demand (Figure 8.6b). The proposed EBQ outperforms the ERA and RRA in terms of energy saving. Meanwhile, the received quality of EBQ is mostly greater than or equal to that of the ERA and RRA. In other words, the EBQ can provide different optimal levels of quality with minimum energy resource consumption, according to the conditions of M-IoT system and video characteristics.

8.5.3.3 EBQ Performance Versus Bandwidth Consumption Coefficient δ_B

To evaluate the performance of EBQ, ERA, and RRA versus bandwidth consumption coefficient δ_B, we fix the quality guarantee coefficient and the skewed

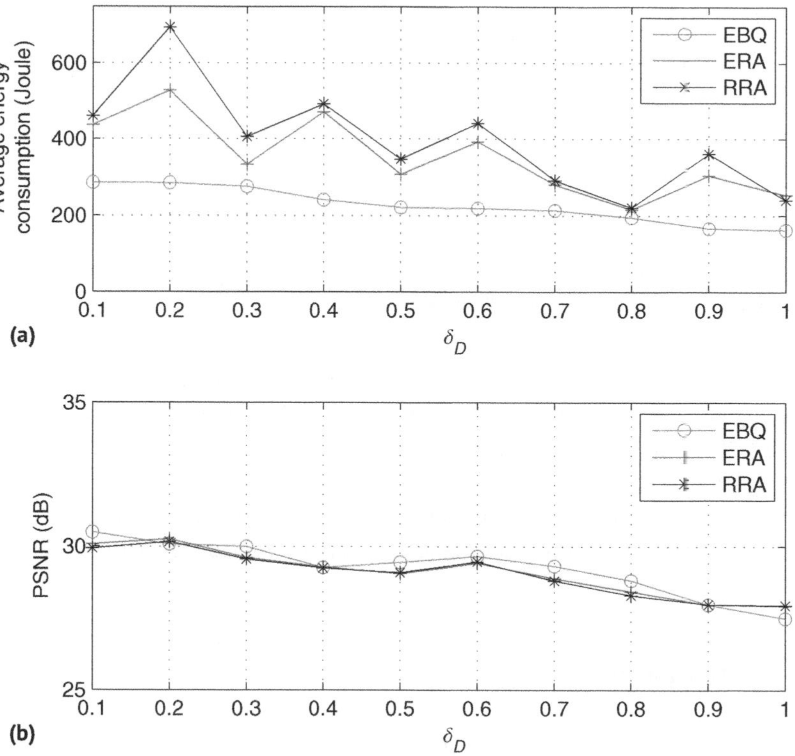

Figure 8.6 Performance evaluation versus δ_D: (a) average energy consumption; (b) PSNR. *Source:* Reproduced from [1] / John Wiley & Sons.

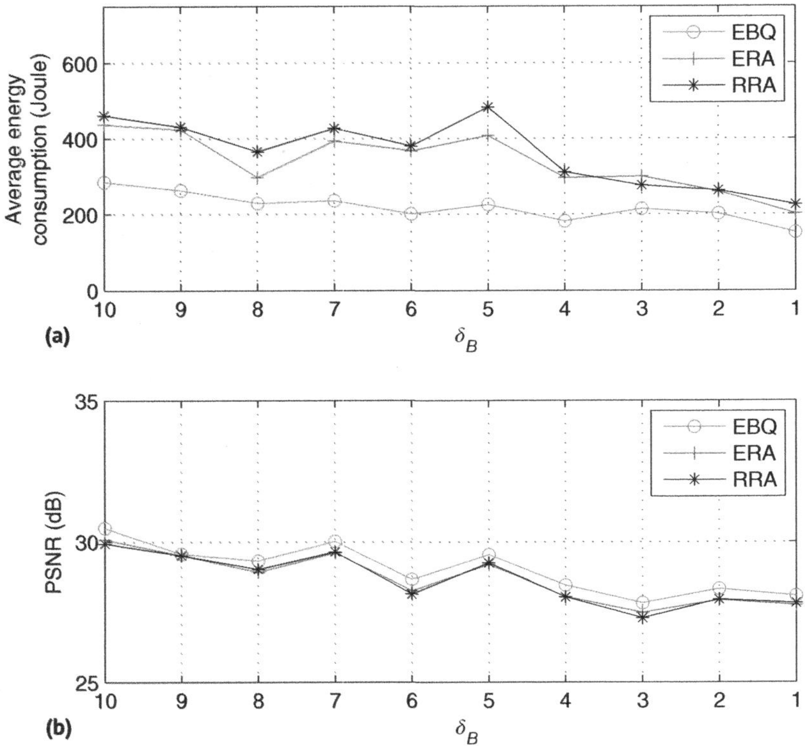

Figure 8.7 Performance evaluation versus δ_B: (a) average energy consumption; (b) PSNR. *Source:* Reproduced from [1] / John Wiley & Sons.

importance parameter as $\delta_D = 0.1$ and $\alpha = 1$, respectively. As shown in Figure 8.7, the average energy consumption (Figure 8.7a) and the quality of received videos (Figure 8.7b) of all schemes decrease with the reduction in bandwidth consumption coefficient, i.e., δ_B decreases, as expected to meet the given backhaul bandwidth constraint of the CHs. In comparison, the EBQ achieves the lowest average energy consumption and provides higher quality of received videos than the ERA and RRA schemes do.

8.6 Conclusion

We have designed the EBQ optimization framework for M-IoT systems to capture video contents and grasp real varying dangers in remote catastrophe-prone areas for better management of natural disasters and rapid urbanization threats. Depending on the conditions of M-IoT systems (i.e., backhaul bandwidth of the CHs and lossy features of the wireless channels) and the videos characteristics (i.e., RDMs and priority/importance of videos), the EBQ adjusted the bandwidth consumption and quality guarantee coefficients and then solved the ARA in EBQ for optimal rates allocated to each video to minimize the average energy consumption of the CSs. Simulation results demonstrated that the EBQ can provide minimum average energy consumption to prolong the lifetime of the CSs while guaranteeing high quality of received videos at the BS and conserving the backhaul bandwidth of the CHs in M-IoT systems.

References

1 N.-S. Vo, D.-B. Ha, B. Canberk, and J. Zhang. Green two-tiered wireless multimedia sensor systems: An energy, bandwidth, and quality optimisation framework. *IET Commun.*, 10(18):2543–2550. 2016.

2 N.-S. Vo, T.-M. Phan, M.-P. Bui, X.-K. Dang, V. T. Nguyen, and C. Yin. Social-aware spectrum sharing and caching helper selection strategy optimized multicast video streaming in dense D2D 5G networks. *IEEE Syst. J.*, 15(3):3480–3491. 2021.

3 M. Alsabah, M. A. Naser, B. M. Mahmmod, S. H. Abdulhussain, M. R. Eissa, and A. Al-Baidhani. 6G wireless communications networks: A comprehensive survey. *IEEE Access*, 9:148191–148243. 2021.

4 Y. Wang, H. Dai, X. Han, P. Wang, Y. Zhang, and C. Xu. Cost-driven data caching in edge-based content delivery networks. *IEEE Trans. Mob. Comput.*, 22(3):1384–1400. 2023.

5 S. Malektaji, A. Ebrahimzadeh, H. Elbiaze, R. H. Glitho, and S. Kianpisheh. Deep reinforcement learning-based content migration for edge content delivery networks with vehicular nodes. *IEEE Trans. Netw. Service Manag.*, 8(3):3415–3431. 2021.

6 K. Duran, E. Ak, G. Yurdakul, and B. Canberk. 6G-enabled DTaaS (Digital Twin as a Service) for decarbonized cities. In *Proc. IEEE Int. Commun. Conf.*, pages 1–6, Rome, Italy, May 2023.

7 G. Kaur, P. Chanak, and M. Bhattacharya. A green hybrid congestion management scheme for IoT-enabled WSNs. *IEEE Trans. Green Commun. Netw.*, 6(4):2144–2155. 2022.

8 R. Chaudhry and N. Kumar. A multi-objective meta-heuristic solution for green computing in software-defined wireless sensor networks. *IEEE Trans. Green Commun. Netw.*, 6(2):1231–1241. 2022.

9 A. Albanese, J. Blomer, J. Edmonds, M. Luby, and M. Sudan. Priority encoding transmission. *IEEE Trans. Inf. Theory*, 42(6):1737–1744. 1996.

10 R. Puri and K. Ramchandran. Multiple description source coding using forward error correction codes. In *Proc. Asilomar Signals, Syst., Comput. Conf.*, pages 342–346, Pacific Grove, CA, Oct. 1999.

11 P. A. Chou, H. J. Wang, and V. N. Padmanabhan. Layered multiple description coding. In *Proc. Packet Video Workshop*, pages 1–7, Nantes, France, Apr. 2003.

12 X. Du, N.-S. Vo, T. Q. Duong, and L. Shu. Joint replication density and rate allocation optimization for VoD systems over wireless mesh networks. *IEEE Trans. Circuits Syst. Video Technol.*, 23(7):1260–1273. 2013.

13 J. M. Boyce, J. Chen, Y. Ye, and A. K. Ramasubramonian. Overview of SHVC: Scalable extensions of the high efficiency video coding standard. *IEEE Trans. Circuits Syst. Video Technol.*, 26(1):20–34. 2016.

14 X. Li, F. Zhou, and J. Du. LDTS: A lightweight and dependable trust system for clustered wireless sensor networks. *IEEE Trans. Inf. Forensics Secur.*, 8(6):924–935. 2013.

15 W. Xiang, C. Zhu, C. K. Siew, Y. Xu, and M. Liu. Forward error correction-based 2-D layered multiple description coding for error-resilient H.264 SVC video transmission. *IEEE Trans. Circuits Syst. Video Technol.*, 19(12):1730–1738. 2009.

16 Y. T. Hou, Y. Shi, and H. D. Sherali. *Applied Optimization Methods for Wireless Networks*. Cambridge University Press, United Kingdom, 1 edition, 2013.

17 R. Zhang, N. Timmons, and J. Morrison. Utility energy-based opportunistic routing for lifetime enhancement in wireless sensor networks. In *Proc. IEEE Int. Commun. Conf.*, pages 6324–6330, London, UK, Jun. 2015.

18 L. Breslau, P. Cao, L. Fan, G. Phillips, and S. Shenker. Web caching and Zipf-like distributions: Evidence and implications. In *Proc. IEEE Int. Conf. on Computer Commun.*, pages 126–134, New York, NY, Mar. 1999.

19 H. Everett. Generalized Lagrange multiplier method for problems of optimum allocation of resources. *Oper. Res.*, 11(13):399–417. 1963.

20 D. E. Goldberg. *Genetic Algorithms in Search, Optimization, and Machine Learning*. Addison-Wesley, Reading, MA, USA, 1988.

21 T. Fang and L. P. Chau. GOP-based channel rate allocation using genetic algorithm for scalable video streaming over error-prone networks. *IEEE Trans. Image Process.*, 15(6):1323–1330. 2006.

22 HEVC. *HM Reference Software Version 12.0*. http://hevc.hhi.fraunhofer.de. 2013.

9

Active Duty Scheduling and Encoding Rate Allocation for Multimedia Applications and Services in Content Delivery Networks

9.1 Introduction

In smart cities, things (e.g., sensors, devices, machines, etc.),[1] citizens, and governments are interconnected in a disruptive platform of information and communication technologies [2]. 6G networks with integrated M-IoT communications play an important role in providing numerous video applications and services (VAS), e.g., surveillance, disaster and environment management, intelligent transportation systems, e-health, and public safety [3]. In M-IoT, CSs capture the video contents from particular areas, packetize and transmit them to the cluster heads (CHs) where they are preprocessed and then forwarded to the BSs for the purpose of monitoring, analyzing, making decisions, and future research.

The greatest challenge is that transmitting videos at high data rate requires huge amounts of bandwidth/throughput and energy resources [4]. Meanwhile, the wireless channels are inherently varying and error-prone, and the CSs are bandwidth- and energy-constrained. Another challenge comes from many videos with high correlation that are captured from the same areas and transmitted to the BS. As a result, more bandwidth and energy resources are wasted on capturing, packetizing, and transmitting the redundant videos. More efforts have been prompted to reduce the resources consumed by VAS while ensuring high quality of service (QoS) in smart cities.

In this chapter, we extend the ARA introduced in Chapter 8 by incorporating ADS, resulting in a joint ADS-ARA model for VAS in M-IoT over 6G networks. The ADS-ARA model consists of two optimization problems. In the ADS optimization problem, the time for each CS on active duty to capture the videos in a cluster is optimally scheduled to minimize the capture energy consumption. The optimal capture times generate the sizes and RDMs of captured videos correspondingly.

---

1 This chapter has been published partly in [1].

In the ARA optimization problem, the captured videos are packetized into multiple descriptions for transmitting by finding the optimal encoding rate. The objective is to minimize the packetization and transmission energy consumption while utilizing the backhaul bandwidth of the BS and ensuring high QoS of received videos.

The ADS-ARA can minimize the capture, packetization, and transmission energy consumption while satisfying given limited bandwidth and high playback quality constraints. Particularly, the ADS optimization problem is formulated for optimally scheduling the active duties of the CSs to capture the videos with minimum capture energy consumption. By taking into account the constraints on bandwidth and quality, we further formulate the ARA optimization problem and solve it for optimally allocating the encoding rates of the captured videos so that they are packetized and transmitted with minimum energy consumption. The other goal of ADS-ARA is that it can mitigate the wasted energy and bandwidth due to the overlap of monitoring the same areas, i.e., capturing, packetizing, and transmitting similar videos to the BS.

Similar to Chapter 8, the RDMs of captured videos are analyzed, and layered multiple description coding with embedded forward error correction (LMDC-FEC) [5–8], based on scalable extensions of high efficiency video coding (SHVC) [9], is employed to packetize the videos into multiple descriptions for transmission. With this approach, the captured videos are packetized and transmitted to the BS while accounting for the diverse system characteristics, including the transmission energy of CSs, video attributes (size, popularity, and RDM), and wireless channel conditions (distance and error), thus significantly improving overall system performance. Simulation results show that the proposed ADS-ARA model for the M-IoT system reduces the capture, packetization, and transmission energy consumption, efficiently utilizes the limited bandwidth, and ensures high playback quality of the received videos at the BS.

The rest of this chapter is organized as follows. The system model is given in Section 9.2. Section 9.3 presents the formulation and analyses of the ADS-ARA model. Section 9.4 describes the ADS-ARA optimization problem and its solution. Simulation results in Section 9.5 demonstrate the benefits of the proposed ADS-ARA solution. Finally, Section 9.6 concludes the chapter.

9.2 ADS-ARA Model

We consider the ADS-ARA model for VAS in M-IoT system over 6G networks as depicted in Figure 9.1. The model consists of a number of M-IoT devices connected to the BS. The devices are grouped into I clusters, each has a cluster head i (CH i) and S_i CSs, $i = 1, 2, \dots, I$. We assume that a lightweight and dependable trust system [10] is applied to elect the CHs from the CSs with more powerful processing

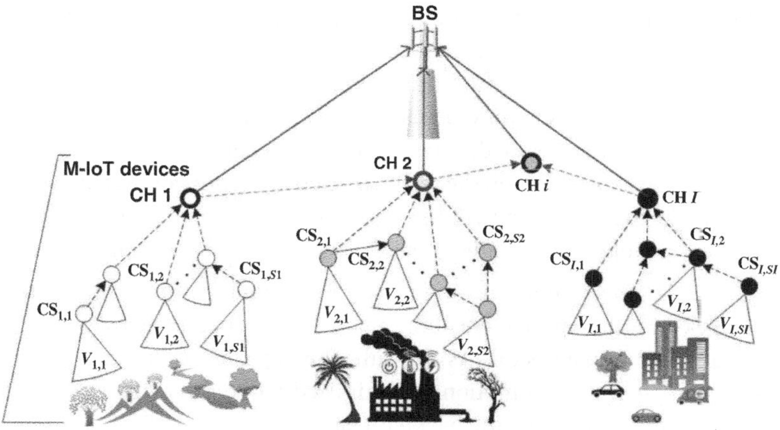

Figure 9.1 ADS-ARA model for VAS in M-IoT over 6G networks.

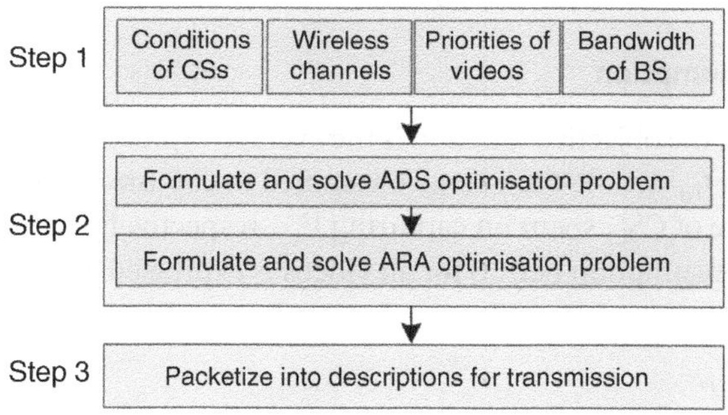

Figure 9.2 ADS-ARA model.

capacity and higher available energy resources. Multi-hop communications are also applied to transmit the captured videos from the CSs to the CHs. CS j in cluster i is identified by $CS_{i,j}$, which is in charge of capturing video $V_{i,j}$, $j = 1, 2, ..., S_i$. Whenever the BS needs the information about the surrounding environments, it requests the CHs and CSs to execute the ADS-ARA model in three steps as shown in Figure 9.2 and detailed as follows:

- Step 1: The BS collects all the system parameters about the conditions of CSs, wireless channels, priorities of videos, and its own backhaul bandwidth limit. The parameters are used for formulating and solving the ADS-ARA optimization problem in step 2.
- Step 2: The BS assigns CH i to formulate and solve the ADS optimization problem for the optimal active duty time $t_{i,j}$ of each $CS_{i,j}$ to capture video $V_{i,j}$. We assume that $V_{i,j}$ is captured at full resolution (maximum encoding rate)

$R_{i,j}$ by manufacturing default configuration. After capturing, $CS_{i,j}$ analyzes $V_{i,j}$ to obtain the size $L^{cap}_{i,j}$ (9.3) and the RDM (i.e., $\gamma_{i,j}$ and $\beta_{i,j}$) (9.2) [11]. Then, all the sizes and RDMs are transmitted to the CHs and subsequently forwarded to the BS, where they are used to formulate the ARA optimization problem. Assume that each video has M layers separated by $M + 1$ encoding rates. Under the constraints on the reconstructed distortion for high playback quality and the backhaul bandwidth of the BS, the ARA optimization problem is solved for the optimal encoding rate $R^{(m)}_{i,j}$ applied to $V_{i,j}$, here $R^{(0)}_{i,j} = 0$ and $m = 1, 2, \ldots, M$. The optimal encoding rate $R^{(m)}_{i,j}$ is sent back to $CS_{i,j}$ for step 3.
- Step 3: Based on the optimal encoding rate $R^{m}_{i,j}$, $CS_{i,j}$ packetizes $V_{i,j}$ into M descriptions for transmission. As a result, the capture, packetization, and transmission energy consumption is minimized while satisfying high playback quality and conserving the backhaul bandwidth at the BS.

9.3 ADS-ARA Formulation and Analysis

9.3.1 Capture Energy Consumption

In ADS optimization problem, the objective function is the average capture energy consumption $\overline{E}_c$. Let $E^{cpb}_{i,j}$ and $t_{i,j}$ denote the energy consumed to capture one bit of $V_{i,j}$ and the active duty time of $CS_{i,j}$ spent on capturing $V_{i,j}$, respectively; then, the average capture energy consumption per CS for all videos is expressed as

$$\overline{E}_c = \sum_{i=1}^{I} \frac{r_i}{S_i} \sum_{j=1}^{S_i} E^{cpb}_{i,j} R_{i,j} t_{i,j}, \tag{9.1}$$

where r_i given in (2.1) [12] represents the importance of a particular area covered by CH i or the access rate of all videos captured by the CSs in cluster i. Here, α features the important distribution of r_i, and it is easy to observe a special case that if $\alpha = 0$, the importance is equally distributed to all clusters and related videos, i.e., $r_i = 1/N, \forall i$.

9.3.2 Packetization and Transmission Energy Consumption

9.3.2.1 Source Video Analysis

Similar to Chapter 8, after capturing, $V_{i,j}$ of size $L^{cap}_{i,j}$ is analyzed to obtain the RDM, which describes the relationship between the encoding rate $R^{(m)}_{i,j}$ and the reconstructed distortion $D_{i,j}(R^{(m)}_{i,j})$, given by [11]

$$D_{i,j}\left(R^{(m)}_{i,j}\right) = \gamma_{i,j}\left(R^{(m)}_{i,j}\right)^{\beta_{i,j}}, \tag{9.2}$$

where $\gamma_{i,j}$ and $\beta_{i,j}$ are chosen to fit the empirical rate–distortion curve of $V_{i,j}$ and $0 \leq R_{i,j}^{(1)} \leq R_{i,j}^{(2)} \leq \dots \leq R_{i,j}^{(m)} \leq \dots \leq R_{i,j}^{(M-1)} \leq R_{i,j}^{(M)} \leq R_{i,j}$.

To packetize $V_{i,j}$ into M descriptions, we apply LMDC-FEC-based SHVC [5–9]. Particularly, $V_{i,j}$ is partitioned into M layers. Layer m, $m = 1, 2, \dots, M$, which is enclosed by two adjacent rates $R_{i,j}^{(m-1)}$ and $R_{i,j}^{(m)}$, is equally divided into m segments. The size of segment m is expressed as [13]

$$l_{i,j}^{(m)}\left(R_{i,j}^{(m)}\right) = \frac{L_{i,j}^{\text{cap}}\left(R_{i,j}^{(m)} - R_{i,j}^{(m-1)}\right)}{mR_{i,j}^{(M)}}. \tag{9.3}$$

M layers of $V_{i,j}$ are protected against erroneous transmission by using Reed–Solomon (RS) code. To this end, in layer m, m original blocks of size $m \times l_{i,j}^{(m)}(R_{i,j}^{(m)})$ are embedded with FEC to create M blocks of size $M \times l_{i,j}^{(m)}(R_{i,j}^{(m)})$, namely RS($M \times l_{i,j}^{(m)}(R_{i,j}^{(m)})$, $m \times l_{i,j}^{(m)}(R_{i,j}^{(m)})$) or LMDC-FEC m. Finally, the m-th blocks extracted from every LMDC-FEC are packetized into description m of size $\sum_{m=1}^{M} l_{i,j}^{(m)}(R_{i,j}^{(m)})$. As a result, $V_{i,j}$ is packetized into M descriptions of total size given by

$$L_{i,j}^{\text{FEC}}\left(R_{i,j}^{(m)}\right) = M \sum_{m=1}^{M} l_{i,j}^{(m)}\left(R_{i,j}^{(m)}\right). \tag{9.4}$$

By applying LMDC-FEC, if m out of M descriptions of $V_{i,j}$ are correctly received at the BS, the video is played back at rate $R_{i,j}^{(m)}$, corresponding to the reconstructed distortion $D_{i,j}(R_{i,j}^{(m)})$, regardless of the order in which the descriptions are received.

9.3.2.2 Packetization Energy Consumption

To compute the packetization energy consumption $\overline{E}_p$, we define $E_{i,j}^{\text{ppb}}$ as the energy to packet one bit of $V_{i,j}$. The average energy consumption per CS to packet all videos can be written by

$$\overline{E}_p = \sum_{i=1}^{I} \frac{r_i}{S_i} \sum_{j=1}^{S_i} L_{i,j}^{\text{FEC}}\left(R_{i,j}^{(m)}\right) E_{i,j}^{\text{ppb}}. \tag{9.5}$$

9.3.2.3 Transmission Energy Consumption

To compute the transmission energy consumption $\overline{E}_t$, we further define $H_{i,j}^{\text{C2C}}$ and $E_{i,j,h}^{\text{bph}}$ as the number of hops from $\text{CS}_{i,j}$ to CH i and the energy to transmit one bit of $V_{i,j}$ over hop h, respectively. The average energy per hop to transmit $V_{i,j}$ of size $L_{i,j}^{\mathrm{FEC}}(R_{i,j}^{(m)})$ bits over $H_{i,j}^{\mathrm{C2C}}$ hops is given by

$$\overline{E}_{i,j}^{L_{i,j}^{\mathrm{FEC}}} = \frac{L_{i,j}^{\mathrm{FEC}}\left(R_{i,j}^{(m)}\right)}{H_{i,j}^{\mathrm{C2C}}} \sum_{h=1}^{H_{i,j}^{\mathrm{C2C}}} E_{i,j,h}^{\mathrm{bph}}, \tag{9.6}$$

where $E_{i,j,h}^{\mathrm{bph}} = \theta_1 + \theta_2 (d_{i,j}^{(h)})^{\eta}$, $\theta_1 = 50\mathrm{nJ/bit/hop}$, $\theta_2 = 0.0013\ \mathrm{pJ/bit/m^4/hop}$, $d_{i,j}^{(h)}$ is the distance of hop h measured in meters, and $\eta = 4$ [14].

It is noted that because most energy is consumed by capture, packetization, and transmission of a large number of the CSs having limited resources compared to the CHs, the transmission energy consumption for communications from the CHs to the BS is not considered in the ARA optimization problem. The average energy consumption per CS to transmit all videos is obtained as

$$\overline{E}_t = \sum_{i=1}^{I} \frac{r_i}{S_i} \sum_{j=1}^{S_i} \overline{E}_{i,j}^{L_{i,j}^{\mathrm{FEC}}}. \tag{9.7}$$

9.3.3 Playback Quality

The playback quality is measured by the average reconstructed distortion $\overline{D}_{i,j}$ of $V_{i,j}$ at the BS under erroneous transmission, given by

$$\overline{D}_{i,j} = \sum_{m=1}^{M} p_{i,j}^{(m)} D_{i,j}(R_{i,j}^{(m)}), \tag{9.8}$$

where $p_{i,j}^{(m)}$ is the correct probability to receive m out of M descriptions of $V_{i,j}$ expressed as

$$p_{i,j}^{(m)} = \binom{m}{M} (1 - e_{i,j})^{m} (e_{i,j})^{M-m}. \tag{9.9}$$

In (9.9), $e_{i,j}$ is the error probability to transmit one description of $V_{i,j}$ from $\mathrm{CS}_{i,j}$ to the BS, expressed as

$$e_{i,j} = 1 - \left[\prod_{h=1}^{H_{i,j}^{\mathrm{C2C}}} \left(1 - e_{i,j,h}^{\mathrm{C2C}}\right) \prod_{l=1}^{H_i^{\mathrm{C2B}}} \left(1 - e_{i,l}^{\mathrm{C2B}}\right) \right], \tag{9.10}$$

where $e_{i,j,h}^{\mathrm{C2C}}$ denotes the error probability of transmitting one description over hop h among $H_{i,j}^{\mathrm{C2C}}$ hops from $\mathrm{CS}_{i,j}$ to CH i, and $e_{i,l}^{\mathrm{C2B}}$ denotes the error probability over hop l among H_i^{C2B} hops from CH i to the BS.

Finally, the average reconstructed distortion of received videos at the BS can be obtained from $\overline{D}_{i,j}$ (9.8) as

$$\overline{D} = \sum_{i=1}^{I} \frac{r_i}{S_i} \sum_{j=1}^{S_i} \overline{D}_{i,j}. \tag{9.11}$$

9.3.4 Backhaul Bandwidth Consumption

In ARA optimization problem, the average backhaul bandwidth consumption, which cannot exceed a given threshold of backhaul bandwidth at the BS, is computed as

$$\overline{B} = \sum_{i=1}^{I} \frac{r_i}{S_i} \sum_{j=1}^{S_i} R_{i,j}^{(M)}. \tag{9.12}$$

9.4 ADS-ARA Optimization Design

9.4.1 ADS Problem and Solution

The ADS optimization problem consists of the objective function (9.1) with respect to (w.r.t) the optimization variable $t_{i,j}$ under the two constraints: the total time T_i of cluster i to capture S_i videos, and the lower bound ($T_{i,j}^{\text{lb}}$) and the upper bound ($T_{i,j}^{\text{ub}}$) of times for $\text{CS}_{i,j}$ to capture $V_{i,j}$. In particular, the ADS optimization problem is formulated as

$$\min_{t_{i,j}} \overline{E}_c, \tag{9.13}$$

$$\text{s.t.} \begin{cases} \sum_{j=1}^{S_i} t_{i,j} = \Gamma T_i, & i = 1, 2, \ldots, I, \\ T_{i,j}^{\text{lb}} \leq t_{i,j} \leq T_{i,j}^{\text{ub}}, & i = 1, 2, \ldots, I,\ j = 1, 2, \ldots, S_i, \end{cases}$$

where $0 < \Gamma \leq 1$ is used to flexibly adjust T_i.

The linear programming optimization problem (9.13) can be solved by using a variant of Mehrotra's predictor–corrector algorithm (i.e., primal-dual interior-point method), namely large-scale method [15, 16].

9.4.2 ARA Problem and Solution

The ARA optimization problem consists of the objective function $\overline{E}_{pt} = \overline{E_p} + \overline{E_t}$ (9.5), (9.7) w.r.t the optimization variable $R_{i,j}^{(m)}$ under the constraints on: the condition of rate allocation presented in Section 9.3.2.1, the limit backhaul bandwidth of the BS (B) and the target reconstructed distortion D for conserving the bandwidth consumption and guaranteeing high playback quality of received videos at the BS. The ARA optimization problem is formulated as

$$\min_{R_{i,j}^{(m)}} \overline{E}_{pt}, \tag{9.14}$$

$$\text{s.t.} \begin{cases} 0 \leq R_{i,j}^{(1)} \leq R_{i,j}^{(2)} \leq \ldots \leq R_{i,j}^{(M)} \leq R_{i,j}, i = 1, 2, \ldots, I, j = 1, 2, \ldots, S_i, \\ \overline{B} \leq \delta_B B, \\ \overline{D} \leq \delta_D D, \end{cases}$$

where $\delta_B > 0$ and $\delta_D > 0$ are used to flexibly adjust B and D, respectively.

The above ARA optimization problem can be solved by genetic algorithms (GAs) [13] and the penalty method [17] as presented in the previous chapters.

9.5 Performance Evaluation of ADS-ARA

9.5.1 System Setting

The parameters of the ADS-ARA model and videos are given in Tables 9.1 and 9.2, respectively. Without loss of generality, we assume that all S_i captured videos in cluster i have the same RDM, i.e., $\gamma_{i,1} = \gamma_{i,2} = \gamma_{i,3}$ and $\beta_{i,1} = \beta_{i,2} = \beta_{i,3}$, but different full capture resolutions, i.e., maximum encoding rate $R_{i,j}$. The size $L_{i,j}^{\text{cap}}$ of $V_{i,j}$ is truncated from the full size $L_{i,j}$ according to the active duty time $t_{i,j}$. To obtain the parameters of videos in Table 9.2, we analyze 5 videos: $V_{1,j}$ (Basketballdrill), $V_{2,j}$ (Basketballpass), $V_{3,j}$ (Foreman), $V_{4,j}$ (Traffic), and $V_{5,j}$ (Racehorses), to get their empirical RDMs by using HM reference software version 12.0 [18] and then generate 5 corresponding pairs of $(\gamma_{i,j}, \beta_{i,j})$.

Table 9.1 Parameter settings.

Symbols	Descriptions
I	5 clusters or CHs
M	64 layers or descriptions per each video
S_i	3 CSs in each cluster
$H_{i,j}^{\text{C2C}}, H_i^{\text{C2B}}$	Randomly from 1 to 5 hops
α	From 0 to 2
$\{T_i\}$	{100, 200, 300, 400, 500}sec
$T_{i,j}^{\text{lb}}$	$0.2T_i, \forall j$

Symbols	Descriptions
$T_{i,j}^{\text{ub}}$	$0.8T_i, \forall j$
$E_{i,j}^{\text{cpb}}$	$5 \times [10^{-7}, 10^{-8}, 10^{-9}; 0.8 \times 10^{-7}, 0.8 \times 10^{-8}, 0.8 \times 10^{-9}; 0.6 \times 10^{-7}, 0.6 \times 10^{-8}, 0.6 \times 10^{-9}; 0.4 \times 10^{-7}, 0.4 \times 10^{-8}, 0.4 \times 10^{-9}; 0.2 \times 10^{-7}, 0.2 \times 10^{-8}, 0.2 \times 10^{-9}; 10^{-8}, 10^{-9}, 10^{-10}]$ J/bit
$E_{i,j}^{\text{ppb}}$	All the CSs consume the same energy of 5×10^{-9} J/bit to packetize the videos
$d_{i,j}^{(h)}$	Randomly from 10 to 50 m
$e_{i,j,h}^{\text{C2C}}, e_{i,l}^{\text{C2B}}$	Randomly from 0.0005 to 0.05
B	30,000 Kbps
D	50 (mean squared error)
δ_B	1
δ_D	From 0.1 to 1

Table 9.2 Parameters of videos.

$V_{i,j}$	Pixels|Frames	$\gamma_{i,j} \vert \beta_{i,j}$	$R_{i,j}$(Kbps)	$L_{i,j}$(Kbit)
$V_{1,1}$	832×480|501	6,566e+01|−0.9714	10,000	1,172,340
$V_{1,2}$	832×480|501	6,566e+01|−0.9714	15,000	1,758,510
$V_{1,3}$	832×480|501	6,566e+01|−0.9714	20,000	2,344,680
$V_{2,1}$	352×288|300	9,806e+00|−0.9972	1,000	11,867
$V_{2,2}$	352×288|300	9,806e+00|−0.9972	2,000	23,734
$V_{2,3}$	352×288|300	9,806e+00|−0.9972	3,000	35,600
$V_{3,1}$	2560×1600|150	1,644e+03|−1.0920	50,000	33,382
$V_{3,2}$	2560×1600|150	1,644e+03|−1.0920	100,000	66,763
$V_{3,3}$	2560×1600|150	1,644e+03|−1.0920	170,000	113,496
$V_{4,1}$	832×480|97	1,108e+03|−1.2060	10,000	160,410
$V_{4,2}$	832×480|97	1,108e+03|−1.2060	20,000	320,820
$V_{4,3}$	832×480|97	1,108e+03|−1.2060	28,300	453,960
$V_{5,1}$	416×240|300	7,652e+01|−1.1530	3,000	198,680
$V_{5,2}$	416×240|300	7,652e+01|−1.1530	4,000	264,906
$V_{5,3}$	416×240|300	7,652e+01|−1.1530	5,300	351,000

9.5.2 ADS Performance Evaluation

We evaluate the proposed optimal ADS in comparison with equal ADS (EDS), random ADS (RDS), and full ADS (FDS). In EDS, all CSs have the same $t_{i,j}$; in RDS, each CS has a random value of $t_{i,j}$; and in FDS, all CSs have the same $t_{i,j} = T_i$, i.e., the videos in cluster i are fully captured within the whole time T_i without considering the problem of wasted energy and bandwidth resources caused by capturing the same areas.

The performance of ADS, EDS, RDS, and FDS is first evaluated versus α in the range of 0–2, while keeping $\Gamma = 1$, as shown in Figure 9.3. We can observe that the energy efficiency increases, i.e., less capture energy consumption, w.r.t the increase of α. In comparison, the ADS consumes the least energy, while the FDS performs the worst among the schemes. The reason is that ADS is able to optimally schedule $t_{i,j}$ for each $CS_{i,j}$ depending on (i) the capture energy consumption per bit and capture resolution defined by manufacturing default configuration and (ii) the priority of captured videos r_i in each cluster, to minimize $\overline{E}_c$, but the other EDS, RDS, and FDS are not.

The performance of ADS, EDS, RDS, and FDS is then investigated versus Γ in the range of 0.1–1 while keeping $\alpha = 1$, as illustrated in Figure 9.4. It is easy to see that when Γ increases for a longer amount of capture durations T_i, $\overline{E}_c$ is higher. The ADS consistently outperforms the other schemes, saving up to 79% of energy compared to the FDS.

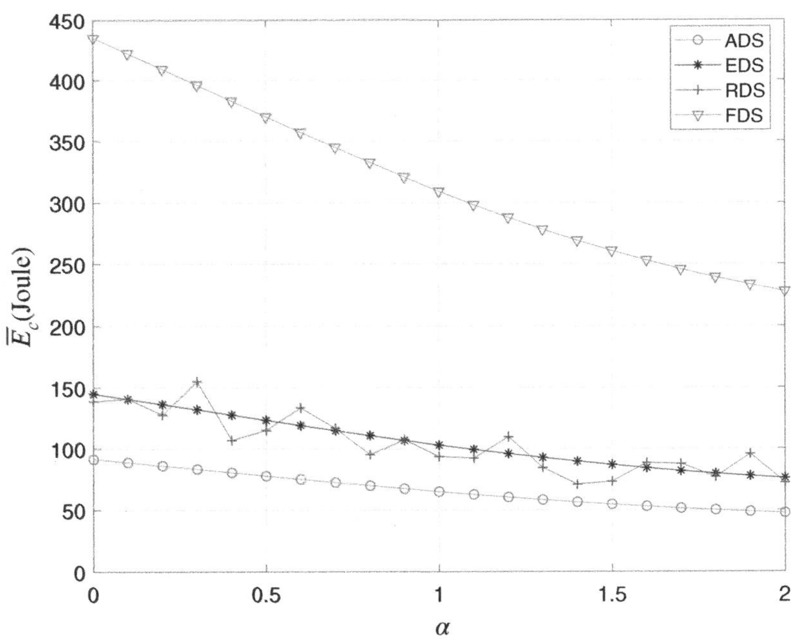

Figure 9.3 Capture energy consumption performance versus α. *Source:* Reproduced from [1] / Springer Nature.

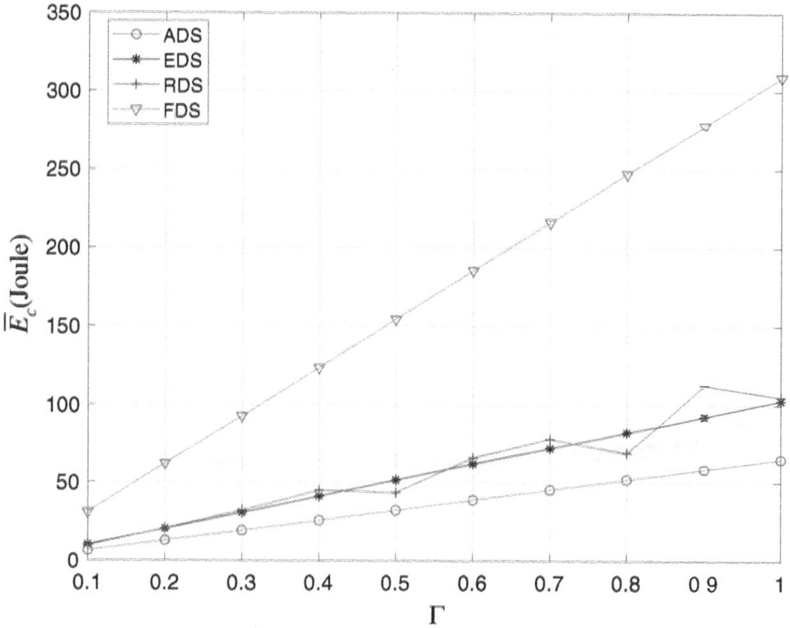

Figure 9.4 Capture energy consumption performance versus Γ. *Source:* Reproduced from [1] / Springer Nature.

9.5.3 ARA Performance Evaluation

To solve the ARA optimization problem, the number of generations of GA is set to 500 in order to ensure the convergence. As shown in Figure 9.5, GA gets converged completely after 400 generations when the mean fitness is equal to the best one, i.e., all individuals have the same best genetic features.

We compare our optimal ARA to the other five schemes, namely, equal ARA (ERA), random ARA (RRA), full equal ARA (FERA), full random ARA (FRRA), and full ARA (FARA). In ERA and RRA, $R_{i,j}^{(m)}$ is equally and randomly allocated in the range of $(0, R_{i,j}^{(M)}]$, respectively. Meanwhile, in FERA and FRRA, the encoding rates are equally and randomly allocated in the range of $(0, R_{i,j}]$, respectively; and in FARA, $R_{i,j}^{(M)}$ is forced to $R_{i,j}$.

9.5.3.1 ARA Performance Versus α

The ARA performance is first evaluated in terms of $\overline{E}_{pt}$, $\overline{D}$ converted into peak signal-to-noise ratio (PSNR) for visual evaluation, and $\overline{B}$, versus α in the range of 0–1 and $\delta_D = 0.5$. We can observe from Figures 9.6–9.8 that the ARA consumes the lowest packetization and transmission energy, i.e., saving up to about

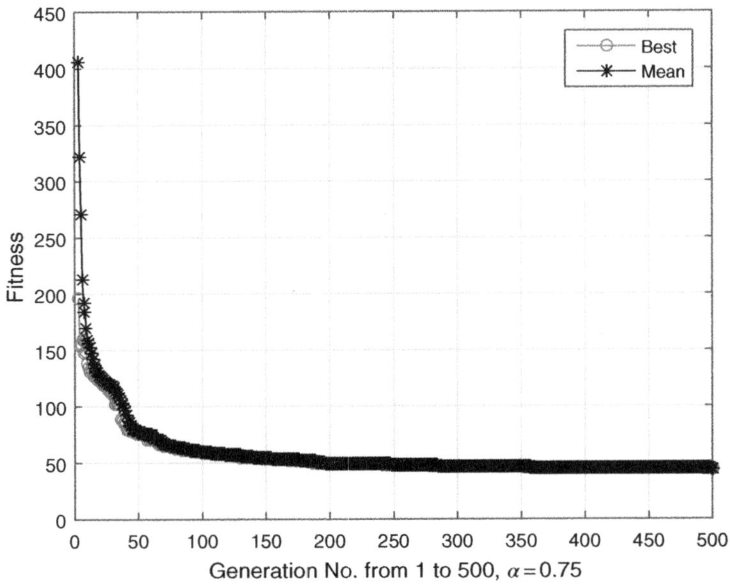

Figure 9.5 Convergence rate of GA. *Source:* Reproduced from [1] / Springer Nature.

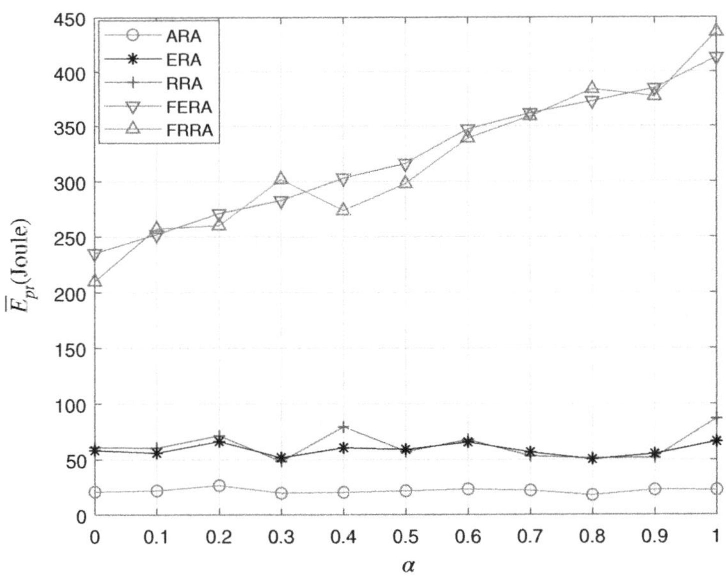

Figure 9.6 Packetization and transmission energy consumption versus α. *Source:* Reproduced from [1] / Springer Nature.

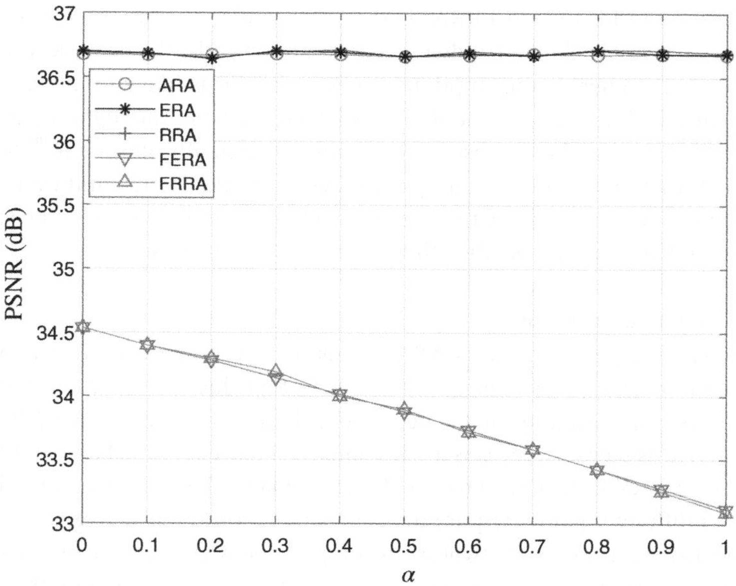

Figure 9.7 Playback quality versus α. *Source:* Reproduced from [1] / Springer Nature.

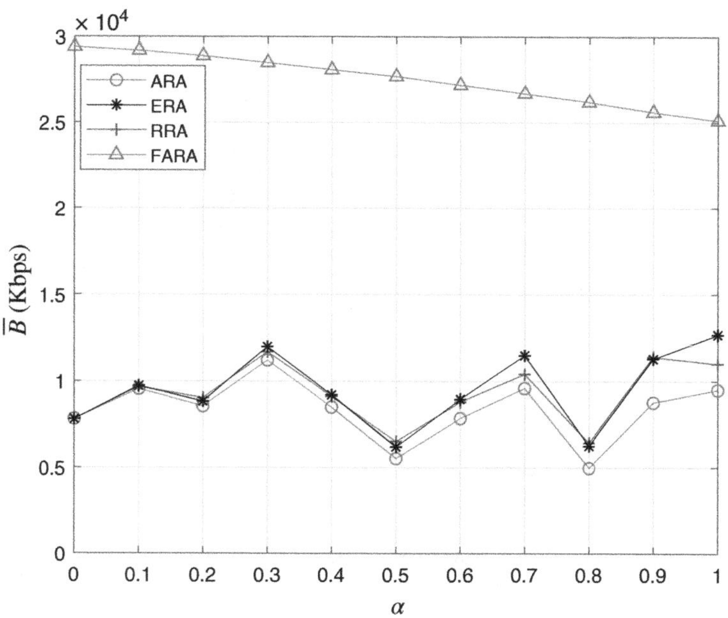

Figure 9.8 Bandwidth consumption versus α. *Source:* Reproduced from [1] / Springer Nature.

95% compared to the FERA and FRRA when $\alpha = 1$, while ensuring high playback quality and lowest backhaul bandwidth consumption at the BS. Especially in Figure 9.6, while the energy consumption of other schemes is likely to increase versus the increase of α, ARA constantly keeps lower to ensure the highest performance. This is because the proposed ARA is able to optimally allocate $R_{i,j}^{(m)}$ to each video, taking into account the energy required for packetization and transmission, the transmission distance, the lossy characteristics of wireless channels, and the video priority; in contrast, the other approaches fail to do so.

9.5.3.2 ARA Performance Versus δ_D

We then evaluate the performance of ARA versus δ_D in the range of 0.1–1 and $\alpha = 0.75$. It can be seen from Figures 9.9 and 9.10 that due to the scarcity of resources, the playback quality of the received videos has to be degraded by increasing δ_D. In this scenario, the CSs use less energy (Figure 9.9) and bandwidth (Figure 9.10) resources to meet a given lower target of quality. That is, the proposed ARA can meet different quality requirements depending on the resource conditions of the M-IoT system in 6G networks. In comparison, ARA gains higher savings on the packetization and transmission energy and backhaul bandwidth than the others.

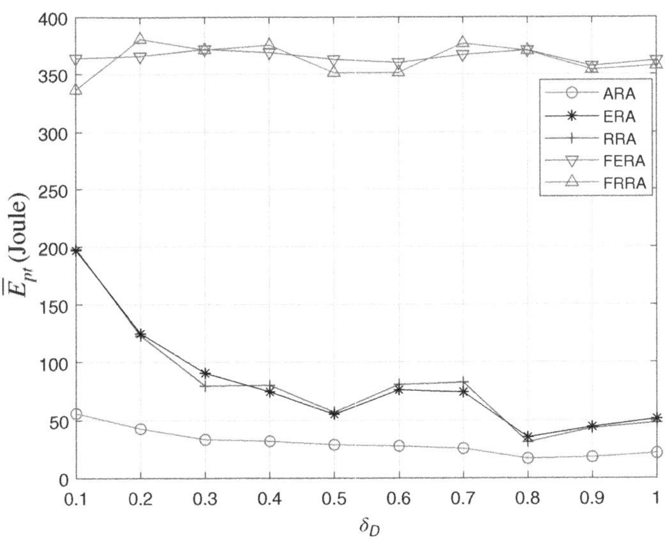

Figure 9.9 Packetization and transmission energy consumption versus δ_D. *Source:* Reproduced from [1] / Springer Nature.

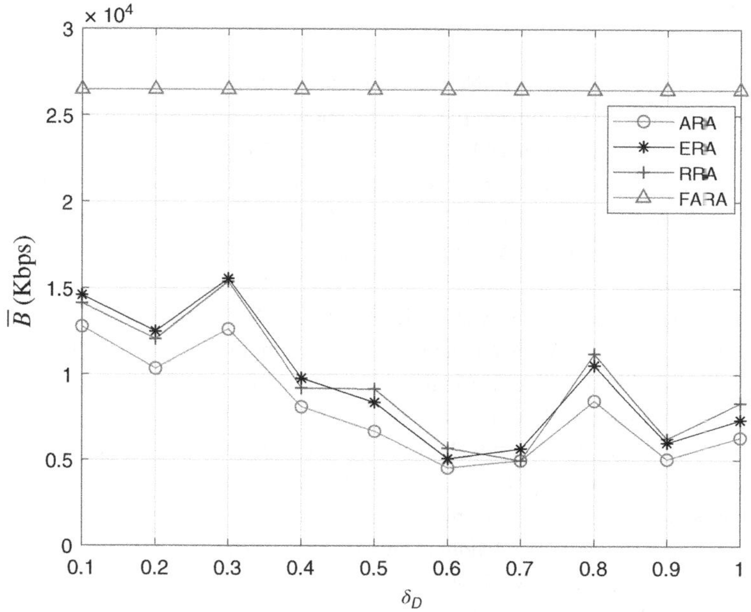

Figure 9.10 Bandwidth consumption versus δ_D. *Source:* Reproduced from [1] / Springer Nature.

9.6 Conclusion

We have proposed an optimization solution for high performance of VAS over M-IoT systems in 6G networks by combining ADS and encoding rate allocation. To do so, many aspects of video (size, RDM, encoding technique, and priority), lossy characteristic, transmission distance, constraints on bandwidth resource and quality, and energy consumption characteristic defined by manufacturing default configuration, were taken into account to find the optimal active duty time and encoding rates of each CS and each video. Consequently, the proposed solution consumed the lowest capture, packetization, and transmission energy while ensuring high playback quality and low backhaul bandwidth consumption at the BS.

References

1 N.-S. Vo, T.-H. Nguyen, and H. K. Nguyen. Joint active duty scheduling and encoding rate allocation optimized performance of wireless multimedia sensor networks in smart cities. *Springer Mobile Netw. Appl.*, 23(6):1586–1596. 2018.

2 R. K. Sethi. *The Role of Telecommunications in Smart Cities*. GlobalLogic Inc., 2015.

3 A. Nauman, Y. A. Qadri, M. Amjad, Y. B. Zikria, M. K. Afzal, and S. W. Kim. Multimedia internet of things: A comprehensive survey. *IEEE Access*, 8:8202–8250. 2020.

4 S. Pratap Singh, N. Kumar, N. S. Alghamdi, G. Dhiman, W. Viriyasitavat, and A. Sapsomboon. Next-gen WSN enabled IoT for consumer electronics in smart city: Elevating quality of service through reinforcement learning-enhanced multi-objective strategies. *IEEE Trans. Consum. Electron.*, 70(4):6507–6518. 2024.

5 A. Albanese, J. Blomer, J. Edmonds, M. Luby, and M. Sudan. Priority encoding transmission. *IEEE Trans. Inf. Theory*, 42(6):1737–1744. 1996.

6 R. Puri and K. Ramchandran. Multiple description source coding using forward error correction codes. In *Proc. Asilomar Signals, Syst., Comput. Conf.*, pages 342–346, Pacific Grove, CA, Oct. 1999.

7 P. A. Chou, H. J. Wang, and V. N. Padmanabhan. Layered multiple description coding. In *Proc. Packet Video Workshop*, pages 1–7, Nantes, France, Apr. 2003.

8 X. Du, N.-S. Vo, T. Q. Duong, and L. Shu. Joint replication density and rate allocation optimization for VoD systems over wireless mesh networks. *IEEE Trans. Circuits Syst. Video Technol.*, 23(7):1260–1273. 2013.

9 J. M. Boyce, J. Chen, Y. Ye, and A. K. Ramasubramonian. Overview of SHVC: Scalable extensions of the high efficiency video coding standard. *IEEE Trans. Circuits Syst. Video Technol.*, 26(1):20–34. 2016.

10 X. Li, F. Zhou, and J. Du. LDTS: A lightweight and dependable trust system for clustered wireless sensor networks. *IEEE Trans. Inf. Forensics Secur.*, 8(6): 924–935. 2013.

11 W. Xiang, C. Zhu, C. K. Siew, Y. Xu, and M. Liu. Forward error correction-based 2-D layered multiple description coding for error-resilient H.264 SVC video transmission. *IEEE Trans. Circuits Syst. Video Technol.*, 19(12):1730–1738. 2009.

12 L. Breslau, P. Cao, L. Fan, G. Phillips, and S. Shenker. Web caching and Zipf-like distributions: Evidence and implications. In *Proc. IEEE Int. Conf. on Computer Commun.*, pages 126–134, New York, NY, Mar. 1999.

13 N.-S. Vo, D.-B. Ha, B. Canberk, and J. Zhang. Green two-tiered wireless multimedia sensor systems: An energy, bandwidth, and quality optimisation framework. *IET Commun.*, 10(18):2543–2550. 2016.

14 Y. T. Hou, Y. Shi, and H. D. Sherali. *Applied Optimization Methods for Wireless Networks*. Cambridge University Press, United Kingdom, 1 edition, 2013.

15 Y. Zhang. Solving large-scale linear programs by interior-point methods under the MATLAB environment. *Optim. Method. Softw.*, 10(1):1–31. 2007.
16 S. Mehrotra. On the implementation of a primal-dual interior point method. *SIAM J. Optim.*, 2:575–601. 1992.
17 T. Fang and L. P. Chau. GOP-based channel rate allocation using genetic algorithm for scalable video streaming over error-prone networks. *IEEE Trans. Image Process.*, 15(6):1323–1330. 2006.
18 HEVC. *HM Reference Software Version 12.0.* http://hevc.hhi.fraunhofer.de. 2013.

10

Optimal Cooperative Video Streaming in Dense D2D Networks

10.1 Introduction

In the previous chapters, to improve the system performance for multimedia applications and services (MAS), many techniques and solutions have been studied: D2D communications, multicasting, caching, resource sharing, power allocation, scheduling, and cooperative transmission.[1] We have also demonstrated the advantages of these techniques and solutions for MAS. However, only Chapters 8 and 9 considered adaptive bitrate (ABR) encoding—the key solution for video streaming. In spite of the benefits of ABR encoding together with active duty scheduling, the two chapters did not propose cooperative transmission solutions to gain better system performance. In addition, ABR encoding was applied to minimize energy consumption, rather than to maximize playback quality, which makes them more suitable for multimedia Internet of Things than for MUs. Another problem is that the quality fluctuation—one of the most important performance metrics of quality of experience (QoE)—has not been modelled and investigated [2, 3].

The emergence of the envisioned future MAS necessitates high-performance, optimal design for D2D communications in both 5G and 6G networks [4–10]. ABR encoding was applied together with resource allocation [4, 7–10], caching [5, 9], mode selection [10], video provider set arrangement and transmission scheduling [6], and cooperative transmission [7]. Although several studies applied ABR encoding, mode selection, and cooperative transmission, their approach relying on the base layer and a few enhancement layers, which must be received in the correct order, lacks adaptability to highly error-prone and time-varying wireless channels [7, 10].

In this chapter, we propose a joint encoding RDO solution for cooperative video streaming (CVS) in dense D2D networks. To achieve the highest data protection against lossy transmission and more fine-grained video streaming under

1 This chapter has been published partly in [1].

varying bandwidth constraints and diverse playback capacities of mobile devices, the layered multiple description coding with embedded forward error correction (LMDC-FEC) method [2, 11, 12] is applied to encode each video segment into multiple descriptions. The RDO problem is formulated for optimal encoding rates allocated to different layers of each video segment and optimal numbers of descriptions distributed to the video helping MUs (VHs) and the MBSs for cooperative transmission. The objective is to minimize the average reconstructed distortion of received video segments for the highest playback quality under the constraints on the energy consumption of VHs and the interference impact on the spectrum helping MUs (SHs). In addition, we investigate the quality fluctuation of received video segments to fulfil the perspective of QoE.

The rest of this chapter is organized as follows. In Section 10.2, we introduce the CVS system with RDO in dense D2D networks. The system is formulated in Section 10.3, allowing us to obtain the constrained nonlinear RDO problem solved by genetic algorithms (GAs) in Section 10.4. Section 10.5 evaluates the performance and demonstrates the benefits of the proposed RDO solution. We conclude the chapter in Section 10.6.

10.2 CVS System with RDO in Dense D2D Networks

We consider a CVS system with RDO in dense D2D networks as shown in Figure 10.1. In this model, the MBS provides network coverage to I requesting users (RUs) and H_i groups of VHs. Each group of VHs, namely GoH, has $n_i^{(h)}$ VHs, including a CH that has already cached the requested video, $i = 1, 2, ..., I$, $h = 1, 2, ..., H_i$ There are also H_i SHs willing to share their spectrum resources

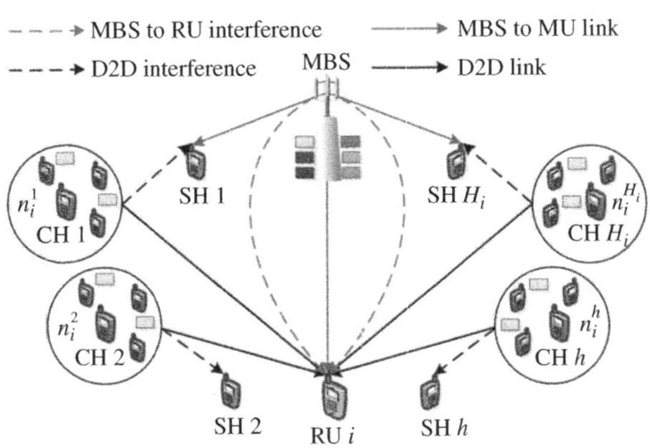

Figure 10.1 A CVS system with RDO in dense D2D networks.

for D2D communications from the CHs in H_i GoHs to RU i. A video of size V is equally divided into S segments, each of size $v = V/S$ bits and then encoded into M descriptions using the LMDC-FEC method based on scalable extensions of high-efficiency video coding [13–15]. An RU is equipped with multiple antennas to receive descriptions simultaneously from the GoHs and the MBS. To deliver all descriptions of the video segments to RU i, the CVS system with RDO operates in three steps as follows:

- Initial step: The MBS collects all necessary information about the cached videos, wireless channels, remaining energy, and so on, then stores them in a stack profile.
- Distribution step: When RU i requests a video, the MBS extracts the information, groups the VHs having similar characteristics of wireless channel to RU i into a GoH, and selects a CH for each GoH. It then formulates and solves the RDO problem for optimally allocating encoding rates to each video segment, packetize each segment into a number of descriptions, and distribute them to GoHs and MBS for cooperative transmission to RU i.
- Update step: The MBS updates the information whenever significant changes occur, replacing outdated data with new one, then stores it for future use.

In the above operation, after solving the RDO problem, the MBS assigns CH h to packetize segment j into M descriptions. Then, GoH h is assigned to send $d_{i,j}^{(h)}$ descriptions. To balance the energy consumption between the VHs in GoH h, CH h sends an equal number of $\frac{d_{i,j}^{(h)}}{n_i^{(h)}}$ to each neighbor for forwarding to RU i. In addition, the spectrum resource shared by SH h for D2D transmission between GoH h and RU i is also reused by the other D2D transmissions between GoH h' and RU i', $i' = 1, 2, \ldots, I, i' \neq i, h' = 1, 2, \ldots, H_i, h' \neq h$.

10.3 RDO Formulation and Analysis

To formulate the RDO problem, we introduce the source rate-distortion model (RDM) and the LMDC-FEC method before transmission. Afterwards, by following the source schemes and taking the characteristic of lossy wireless channel into account, the end-to-end reconstructed distortion expression of received video segments is derived as the objective function of the optimization problem. The energy consumption and co-channel interference are derived as the two important constraints of the RDO problem to save the energy of VHs and guarantee the target signal-to-interference-plus-noise ratio (SINR) of SHs.

10.3.1 RDM and LMDC-FEC Packetization

Consider segment j of the requested video, $j = 1, 2, ..., S$. When m out of M descriptions are correctly received, $m = 1, 2, ..., M$, i.e., first m layers are decoded, the relationship between the reconstructed distortion $D_j(R_j^{(m)})$ and the encoding rate $R_j^{(m)}$ takes the form of the decaying exponential RDM as

$$D_j(R_j^{(m)}) = \gamma_j (R_j^{(m)})^{\beta_j}, \tag{10.1}$$

where γ_j and β_j are the sequence-dependent parameters found to meet the experimental rate–distortion results of standard video sequences [12, 16].

Based on the RDM, segment j is packetized into M descriptions by LMDC-FEC [2, 11, 15, 17], as shown in Figure 10.2. In particular, there are three steps to packetize segment j as follows:

- Step 1—Layering: Segment j is partitioned into M layers. The layers are separated from each other by $M + 1$ encoding rate points $R_j^{(m)}$ measured in Kbps. The rates are allocated such that $0 = R_j^{(0)} \leq R_j^{(1)} \leq R_j^{(2)} \leq ... \leq R_j^{(M)}$. Then, layer m is equally divided into m blocks. In the j-th segment, the sizes of layer m and block m are, respectively, computed as

$$B_j^{(m)} = \frac{\upsilon(R_j^{(m)} - R_j^{(m-1)})}{R_j^{(M)}}, \tag{10.2}$$

$$b_j^{(m)} = \lfloor \frac{B_j^{(m)}}{m} \rfloor, \tag{10.3}$$

  where the floor operator $\lfloor . \rfloor$ is used to round a given value to the nearest integer less than or equal to it. This operator may cause $\sum_{m=1}^{M} m b_j^{(m)} < \upsilon$, but it is infinitesimal and does not have a serious impact on the packetization process. For convenience, we, therefore, do not dwell on a bit adjustment algorithm to cope with the bits truncated by the floor operator.
- Step 2—Embedding FEC: By using Reed–Solomon (RS) code, an RS $(M \times b_j^{(m)}, m \times b_j^{(m)})$ is applied across m blocks of size $m \times b_j^{(m)}$ bits of layer m to create M blocks of size $M \times b_j^{(m)}$ bits with embedded FEC, namely layer m with embedded FEC. So far, however, receiving m layers correctly cannot ensure that segment j is encoded to playback up to $R_j^{(m)}$ Kbps if the first m layers are not correctly received in order.
- Step 3—Packetizing: To remove the order among the received layers, description m is packetized by selecting the m-th blocks from M layers with embedded FEC. Hence, we obtain M descriptions with the same size of $\sum_{m=1}^{M} b_j^{(m)}$ bits. The total size of segment j with embedded FEC is given by

$$\upsilon_j^{\text{FEC}} = M \sum_{m=1}^{M} b_j^{(m)}. \tag{10.4}$$

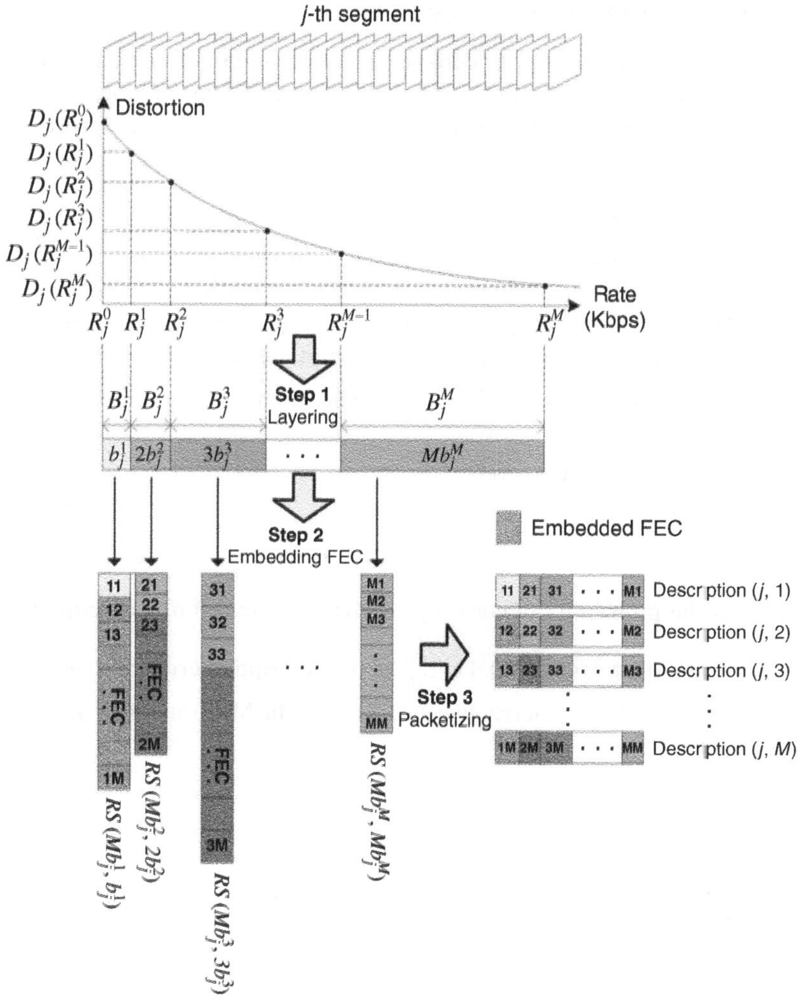

Figure 10.2 LMDC-FEC packetization model. *Source:* [1]/IEEE/CC BY 4.0.

With embedded FEC, the RS $\left(M \times b_j^{(m)}, m \times b_j^{(m)}\right)$ can correct up to $(M - m)$ erroneous blocks of the m-th layer. Therefore, if there are m out of M descriptions correctly received, the first m layers of segment j are decoded and it is played back at the reconstructed quality corresponding to rate $R_j^{(m)}$. The reconstructed quality of received segments depends on only the number of descriptions but not the order of layers correctly received.

10.3.2 End-to-end Reconstructed Distortion

We define $P_{i,j}^{(m)}$ as the probability of correctly receiving m out of M descriptions of segment j. Based on the RDM (10.1), by taking into account the intra-access rate r_j of segment j, the average reconstructed distortion of received video with S segments is given by

$$\overline{D} = \frac{1}{I}\sum_{i=1}^{I}\sum_{j=1}^{S} r_j\gamma_j\left[\sum_{m=0}^{M} P_{i,j}^{(m)} R_j^{(m)}\right]^{\beta_j}. \tag{10.5}$$

In (10.5), M descriptions of segment j are sent to RU i by H_i GoHs and the MBS. GoH h is responsible for sending $d_{i,j}^{(h)}$ descriptions, while the MBS sends the remaining ones, i.e., $d_{i,j}^{(0)} = M - \sum_{h=1}^{H_i} d_{i,j}^{(h)}$. In addition, $P_{i,j}^{(m)}$ is calculated as

$$P_{i,j}^{(m)} = \sum_{\substack{\{m_h\}:\sum_{h=0}^{H_i} m_h = m \\ m_h \le d_{i,j}^{(h)}}} \prod_{h=0}^{H_i} P_{d_{i,j}^{(h)}}^{(m_h)}, \tag{10.6}$$

where $P_{d_{i,j}^{(h)}}^{(m_h)}$ is the probability of correctly receiving m_h out of $d_{i,j}^{(h)}$ descriptions sent by GoH h and the MBS ($h = 0$). Let $\varepsilon_{i,j}^{(h)}$ be the description error rate over the common channel (or the channel) between GoH h (or the MBS) and RU i, $P_{d_{i,j}^{(h)}}^{(m_h)}$ is given by

$$P_{d_{i,j}^{(h)}}^{(m_h)} = \begin{pmatrix} m_h \\ d_{i,j}^{(h)} \end{pmatrix}\left(1-\varepsilon_{i,j}^{(h)}\right)^{m_h}\left(\varepsilon_{i,j}^{(h)}\right)^{d_{i,j}^{(h)}-m_h}. \tag{10.7}$$

For example, in (10.6), if $m = 0$ (i.e., $m_h = 0, h = 0, 1, 2, \dots, H_i$), we have

$$P_{i,j}^{(0)} = P_{d_{i,j}^{(0)}}^{(0)} \times P_{d_{i,j}^{(1)}}^{(0)} \times P_{d_{i,j}^{(2)}}^{(0)} \times \dots \times P_{d_{i,j}^{(H_i-1)}}^{(0)} \times P_{d_{i,j}^{(H_i)}}^{(0)}, \tag{10.8}$$

and if $m = 1$ (i.e., $m_h = 1; \{m_k\} = 0, k = 0, 1, 2, \dots, H_i, k \neq h$), we obtain

$$\begin{aligned} P_{i,j}^{(1)} =& P_{d_{i,j}^{(0)}}^{(0)} \times P_{d_{i,j}^{(1)}}^{(0)} \times P_{d_{i,j}^{(2)}}^{(0)} \times \dots \times P_{d_{i,j}^{(H_i-1)}}^{(0)} \times P_{d_{i,j}^{(H_i)}}^{(1)} + \\ & P_{d_{i,j}^{(0)}}^{(0)} \times P_{d_{i,j}^{(1)}}^{(0)} \times P_{d_{i,j}^{(2)}}^{(0)} \times \dots \times P_{d_{i,j}^{(H_i-1)}}^{(1)} \times P_{d_{i,j}^{(H_i)}}^{(0)} + \\ & \cdots + \\ & P_{d_{i,j}^{(0)}}^{(1)} \times P_{d_{i,j}^{(1)}}^{(0)} \times P_{d_{i,j}^{(2)}}^{(0)} \times \dots \times P_{d_{i,j}^{(H_i-1)}}^{(0)} \times P_{d_{i,j}^{(H_i)}}^{(0)}, \end{aligned} \tag{10.9}$$

and so on, if $m = M$ (i.e., $m_h = d_{i,j}^{(h)}$), we have

$$P_{i,j}^{(M)} = P_{d_{i,j}^{(0)}}^{d_{i,j}^{(0)}} \times P_{d_{i,j}^{(1)}}^{d_{i,j}^{(1)}} \times P_{d_{i,j}^{(2)}}^{d_{i,j}^{(2)}} \times \ldots \times P_{d_{i,j}^{(H_i-1)}}^{d_{i,j}^{(H_i-1)}} \times P_{d_{i,j}^{(H_i)}}^{d_{i,j}^{(H_i)}}. \tag{10.10}$$

For the sake of computational implementation, (10.5) can be rewritten in the form of

$$\overline{D} = \frac{1}{I}\sum_{i=1}^{I}\sum_{j=1}^{S} r_j \gamma_j \left[\sum_{m_0=0}^{d_{i,j}^{(0)}} P_{d_{i,j}^{(0)}}^{(m_0)} \sum_{m_1=0}^{d_{i,j}^{(1)}} P_{d_{i,j}^{(1)}}^{(m_1)} \sum_{m_2=0}^{d_{i,j}^{(2)}} P_{d_{i,j}^{(2)}}^{(m_2)} \right.$$

$$\left. \cdots \sum_{m_{H_i-1}=0}^{d_{i,j}^{(H_i-1)}} P_{d_{i,j}^{(H_i-1)}}^{(m_{H_i-1})} \sum_{m_{H_i}=0}^{d_{i,j}^{(H_i)}} P_{d_{i,j}^{(H_i)}}^{(m_{H_i})} R_j^{(\sum_{h=0}^{H_i} m_h)} \right]^{\beta_j}. \tag{10.11}$$

In (10.11), the intra-access rate r_j represents the popularity of segment j such that $\sum_{j=1}^{S} r_j = 1$. Statistical studies show that different intra-access rates of segments are often defined by Zipf-like distribution [18] given by

$$r_j = \frac{j^{-\alpha}}{\sum_{j'=1}^{S} (j')^{-\alpha}}, \tag{10.12}$$

where α is the skewed access rate coefficient used to reflect the popularity pattern of segments. For example, $\alpha = 0$ means that all segments have the same intra-access rate of $1/S$, while higher values of α make the intra-access rates much more different between the segments.

We can observe from (10.11) that $\overline{D}$ can be minimized for high-quality playback by optimally allocating $R_j^{\left(\sum_{h=0}^{H_i} m_h\right)}$ (or actually $R_j^{(m)}$) and $d_{i,j}^{(h)}$ in accordance with $P_{d_{i,j}^{(h)}}^{(m_h)}$ and r_j. Furthermore, the optimal values of $R_j^{(M)}$, i.e., the maximum encoding rate, and $d_{i,j}^{(h)}$ can be found such that they satisfy the constraints of average energy consumption of all VHs and the target SINR of SHs for energy savings and proper co-channel interference impact, respectively. These two constraints are presented in the following subsections.

10.3.3 Energy Consumption

The energy consumption of GoH h for sending $d_{i,j}^{(h)}$ descriptions of segment j to RU i is given by

$$E_{i,j}^{(h)} = t_{i,j}^{(h)} P_{i,j}^{(h)}, \tag{10.13}$$

where the time $t_{i,j}^{(h)}$ is

$$t_{i,j}^{(h)} = \frac{d_{i,j}^{(h)} \sum_{m=1}^{M} b_j^{(m)}}{10^3 R_j^{(M)}}, \tag{10.14}$$

and $P_{i,j}^{(h)}$ is the common transmission power of GoH h and the transmission power of the MBS ($h = 0$). Given the transmission rate $R_j^{(M)}$, by assuming the Gaussian channel with Shannon-like capacity between GoH h/the MBS and RU i [19, 20], we have

$$P_{i,j}^{(h)} = \begin{cases} \left(G_{i,j}^{(h)}\right)^{-1} \left(2^{\frac{R_j^{(M)}}{W}} - 1\right)\left(P_{\mathrm{SH},i}^{(0,h)} G_{\mathrm{SH},i}^{(0,h)} + \sum\limits_{\substack{i'=1 \\ i' \neq i}}^{I} \sum\limits_{\substack{h'=1 \\ h' \neq h}}^{H_i} P_{i',j}^{(h')} G_{i',j}^{(h')} + \sigma_0\right), h \geq 1, \\ \left(G_{i,j}^{(0)}\right)^{-1} \left(2^{\frac{R_j^{(M)}}{W}} - 1\right)\sigma_0, h = 0, \end{cases} \tag{10.15}$$

where W is the system bandwidth, $G_{i,j}^{(h)}$ and $G_{i,j}^{(0)}$ are the channel gains from GoH h and the MBS to RU i related to segment j, $P_{\mathrm{SH},i}^{(0,h)}$ and $G_{\mathrm{SH},i}^{(0,h)}$ are the transmission power and the channel gain from the MBS to SH h related to RU i, and σ_0 is the additive white Gaussian noise (AWGN) power. The channel gain $G_{i,j}^{(h)}$, which is assumed unchanged during the transmission of all M descriptions of segment j, is represented by the path loss $g_{i,j}^{(h)}$ and Rayleigh fading $|c_{i,j}^{(h)}|^2$ with $\mathbb{E}[|c_{i,j}^{(h)}|^2] = 1$, as follows:

$$G_{i,j}^{(h)} = g_{i,j}^{(h)} |c_{i,j}^{(h)}|^2. \tag{10.16}$$

Finally, the overall average energy consumed per VH to send S segments to I RUs can be expressed as

$$\overline{E} = \sum_{i=1}^{I} \sum_{j=1}^{S} \frac{r_j}{H_i} \sum_{h=1}^{H_i} \frac{E_{i,j}^{(h)}}{n_i^{(h)}}. \tag{10.17}$$

In (10.17), we ignore the energy consumed by the MBS because the available energy of the MBS is extremely higher than that of the VHs.

10.3.4 SINR at SHs

When SH h shares its spectrum resource for D2D transmission from GoH h to RU i, it is also affected by the interference power $P_{i',j}^{(h')}$ associated with simultaneous transmissions from other GoHs $h' \neq h$ to their respective RUs $i' \neq i$. Accordingly, the SINR at SH h is computed as

$$\gamma_{\text{SH},i}^{(0,h)} = \frac{P_{\text{SH},i}^{(0,h)} G_{\text{SH},i}^{(0,h)}}{\sum_{\substack{i'=1 \\ i' \neq i}}^{I} \sum_{\substack{h'=1 \\ h' \neq h}}^{H_i} P_{i',j}^{(h')} G_{i',j}^{(h')} + \sigma_0}. \tag{10.18}$$

We derive (10.18) to guarantee the SINR requirement at SH h when sharing its spectrum resource.

10.4 RDO Optimization Design

10.4.1 RDO Problem

The objective of the RDO problem is to minimize the average reconstructed distortion of received segments for high playback quality by finding the optimal matrices of encoding rates $\mathbf{R}_{S \times M} = \{R_j^{(m)}\}$ and the number of descriptions $\mathbf{d}_{S \times I \times H_i} = \{d_{i,j}^{(h)}\}$. Taking into account the constraints of encoding rates, video and segment sizes, number of descriptions of segment j transmitted to RU i, remaining energy of VHs (E), and target SINR of SHs (γ_0), the RDO problem can be formulated as follows:

$$\min_{\mathbf{R},\mathbf{d}} \overline{D} \tag{10.19}$$

$$\text{s.t.} \begin{cases} 0 = R_j^{(0)} \leq R_j^{(1)} \leq \ldots \leq R_j^{(M)},\ j = 1, 2, \ldots, S,\ \sum_{j=1}^{S} \sum_{m=1}^{M} B_j^{(m)} \leq \delta V, \\ \sum_{m=1}^{M} B_j^{(m)} \leq \upsilon,\ j = 1, 2, \ldots, S, \\ \sum_{h=0}^{H_i} d_{i,j}^{(h)} \leq M,\ i = 1, 2, \ldots, I,\ j = 1, 2, \ldots, S, \\ \overline{E} \leq E, \\ \sum_{\substack{i'=1 \\ i' \neq i}}^{I} \sum_{\substack{h'=1 \\ h' \neq h}}^{H_i} P_{i',j}^{(h')} G_{i',j}^{(h')} \leq \dfrac{P_{\text{SH},i}^{(0,h)} G_{\text{SH},i}^{(0,h)}}{\gamma^{(0)}} - \sigma_0, \\ i = 1, 2, \ldots, I,\ j = 1, 2, \ldots, S,\ h = 1, 2, \ldots, H_i, \end{cases} \tag{10.20}$$

where $0 < \delta \leq 1$ is used to adjust the video size to meet the processing capacity and resolution of RUs. The last constraint, which is to limit the interference impact on the SHs, is obtained from (10.18) by letting $\gamma_{\mathrm{SH},i}^{(0,h)} \geq \gamma^{(0)}$.

The proposed RDO problem is very complicated to solve by Lagrange optimization approach or even traditional methods such as Karush–Kuhn–Tucker [2] because we cannot compute the gradient terms with respect to $\mathbf{d}_{S\times I\times H_i}$ occurring as the upper limits of summations in the objective function. In practice, search methods to obtain an exact or approximate solution with reasonable memory and time complexity are preferable to that with global optimum but excessive complexity. In this regard, GAs—adaptive heuristic search algorithms for exact or approximate optimal solution based on the evolutionary principles of nature selection and genetic variation—can be applied to solve (10.19) and (10.20). Whilst conventional search approaches may result in a local optimum if the search space is multimodal, GA can utilize random property to avoid being trapped in local optimum and take into account many optimal peaks in parallel, despite unimodal or multimodal search space [2, 21]. The GA implementation is described in the sequel.

10.4.2 GA Solution

The GA solution can deal with simple constraints in the form of lower and upper bounds but not complicated ones such as (10.20). To address the complicated constraints, we use the penalty function [21], which requires to convert (10.19) and (10.20) to an unconstrained optimization problem. To do so, all the constraints in (10.20) are rewritten as

$$
\begin{cases}
\triangle R_j^{(m)} = R_j^{(m+1)} - R_j^{(m)} \geq 0, m = 1, 2, \dots, M-1, \\
\triangle V = \delta V - \sum_{j=1}^{S} \sum_{m=1}^{M} B_j^{(m)} \geq 0, \\
\triangle v_j = v - \sum_{m=1}^{M} B_j^{(m)} \geq 0, \\
\triangle d_{i,j} = M - \sum_{h=0}^{H_i} d_{i,j}^{(h)} \geq 0, \\
\triangle E = E - \overline{E} \geq 0, \\
\triangle P_{i,j}^{(h)} = \dfrac{P_{\mathrm{SH},i}^{(0,h)} G_{\mathrm{SH},i}^{(0,h)}}{\gamma^{(0)}} - \sigma_0 - \sum_{\substack{i'=1 \\ i' \neq i}}^{I} \sum_{\substack{h'=1 \\ h' \neq h}}^{H_i} P_{i',j}^{(h')} G_{i',j}^{(h')} \geq 0.
\end{cases}
\tag{10.21}
$$

We obtain the penalty function with respect to all the above constraints as

$$\begin{aligned} F = {} & \lambda_1 \sum_{j=1}^{S} \sum_{m=1}^{M-1} [\min\{0, \triangle R_j^{(m)}\}]^2 \\ & + \lambda_2[\min\{0, \triangle V\}]^2 + \lambda_3 \sum_{j=1}^{S}[\min\{0, \triangle \upsilon_j\}]^2 \\ & + \lambda_4 \sum_{i=1}^{I} \sum_{j=1}^{S}[\min\{0, \triangle d_{i,j}\}]^2 + \lambda_5[\min\{0, \triangle E\}]^2 \\ & + \lambda_6 \sum_{i=1}^{I} \sum_{j=1}^{S} \sum_{h=0}^{H_i}[\min\{0, \triangle P_{i,j}^{(h)}\}]^2, \end{aligned} \tag{10.22}$$

where λs, i.e., λ_1, λ_2, λ_3, λ_4, λ_5, and λ_6, are the degree of constraint violation. To obtain the values of λs, we carefully increase them initially from small values to large values [21]. At the beginning, we set them to 10^{-3} and check F. If F is large and all elements in (10.22) associated to λs are not equal to zero, we increase λs by the factor of 10. A λ is kept not to increase if its corresponding element is equal to zero. The process is repeated until $F = 0$. In the final steps, the factors are more fine-grained, i.e., 5 and 2 instead of 10, to ensure that they do not impact on the other elements with λs fixed in the previous steps.

Finally, by combining (10.19) and (10.22), we have the following unconstrained optimization problem:

$$\min_{\mathbf{R},\mathbf{d}} \overline{D}_F = \overline{D} + F. \tag{10.23}$$

So far, instead of solving (10.19) and (10.20), we implement GA to solve (10.23) for optimal results of **R** and **d**. However, in GA, the optimal results are generated in an array instead of in a matrix for the ease of Gray binary coding. Therefore, $\mathbf{R}_{S\times M}$ and $\mathbf{d}_{S\times I\times H_i}$ are hereafter represented by $\mathbf{R}'_{1\times SM}$ and $\mathbf{d}'_{1\times SIH_i}$, or $\mathbf{R}'$ and $\mathbf{d}'$ for simplicity, respectively. The implementation of GA is detailed in Algorithm 10.1, whose memory complexity is $\mathcal{O}(N_\mathrm{P}N_\mathrm{B})$ and time complexity depends on the number of generations (N_G). In each generation, the time complexity further depends on the population size (N_P), the number of bits representing an individual (N_B), the genetic operators (i.e., selection, crossover, and mutation), and the fitness evaluation. Here, the time complexity of mutation can be neglected because it is rarely done with $P_\mathrm{M} = 10^{-6}$. It is noted that the RUs do not have to wait too long before getting the first frames for playing back thanks to the segment by segment packetization process and the powerful processing capacity of the MBS to implement the GA.

Algorithm 10.1 Implementation of GA

Input: Parameters listed in Table 10.1
$N_{\mathrm{P}} = 500$: Population size, i.e., number of individuals
$N_{\mathrm{B}} = M + H_i + 1$: Number of bits to represent an individual of solution set $\{C_k\} = [\mathbf{R}'_k, \mathbf{d}'_k]$, $k = 1, 2, \ldots, N_{\mathrm{P}}$, it is equivalent to the number of rates
$N_{\mathrm{G}} = 500$: Number of generations
$P_{\mathrm{G}} = 0.9$: Generation gap
$P_{\mathrm{C}} = 0.9$: Crossover probability
$P_{\mathrm{M}} = 10^{-6}$: Mutation probability
Output: $C^* = [\mathbf{R}'^*, \mathbf{d}'^*]$

```
Generating N_P random individuals of solution {C_k}
Calculating fitness values D̄_F(C_k) in (10.23) corresponding to each individual in {C_k} of current generation
while N_G > 0 do
    Putting {C_k} and D̄_F({C_k}) in mating pool
    Selecting N_P × P_G best individuals with better fitness values, i.e., lower values of D̄_F(C_k), for breeding next generation by applying stochastic universal sampling operator [22]
    Continuing to select 2 parent individuals to generate 2 offspring by using single point crossover operator [23] with crossover probability P_C
    Mutating 2 new offspring with mutation probability P_M to get back the good genetic characteristics that could be lost in the previous operators
    Evaluating fitness values of all offspring and reinserting them into the current population
    N_G ← N_G − 1
end while
Finding the best fitness value and the best corresponding individual C* in the last generation
```

10.5 Performance Evaluation of RDO

10.5.1 System Setting

The system's parameters are given in Table 10.1. Without loss of generality, the system consists of 1 RU, 1 video (namely Foreman sequence) divided into 8 segments, 5 GoHs, and 10 VHs per GoH. The distance from the MBS to the RU is less than 1 km in urban areas, and it is set to 300 m especially in dense D2D networks. The distance from the VHs to the RU is randomly distributed from 50 m to 100 m. Correspondingly, the maximum powers of MBS and VHs are 15 W and 0.5 W, respectively. The description error rate is randomly distributed from 0.05 to 0.5, and the system bandwidth is 5 MHz. We assume that all the segments of Foreman sequence have similar RDM.

10.5.2 Convergence Evaluation of GA

The convergence rate of GA is shown in Figure 10.3. The convergence holds after about 25 generations, which clearly indicates the feasibility of using GA in solving the RDO problem. The best and average fitness values of all individuals gradually

Table 10.1 Parameters setting.

Symbols	Specifications
I	1 RU
$H_i = H$	1–5 GoHs
$n_i^{(h)} = n$	10 VHs per GoH
S	8 segments of CIF (352 × 288) Foreman sequence
υ	50 KB
M	64 layers (or descriptions) of a segment
$\{\gamma_j, \beta_j\}$	$\{1.914 \times 10^4, -1.20515\}$ [12, 16]
$\varepsilon_{i,j}^{(h)}$	Randomly distributed from 0.05 to 0.5 [2, 3, 21]
δ	75% of full size of Foreman sequence
E	2.5–100 Joules [24, 25]
W	5 MHz
σ_0	10^{-13} W [3, 26]
α	0.75 [18]
$\{\lambda_{1,2,3,4,5,6}\}$	{1, 0.1, 0.1, 500, 500, 0.1}

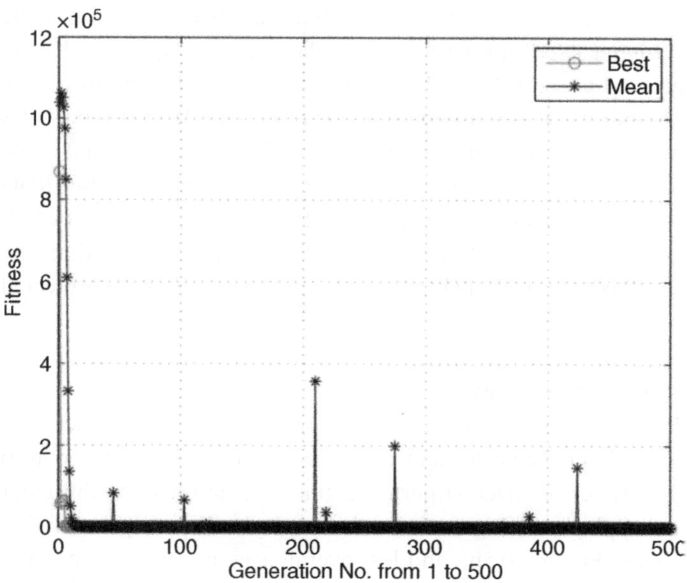

Figure 10.3 Convergence performance of GA. *Source:* [1]/IEEE/CC BY 4.0.

improve and converge from generation to generation, following the principles of natural evolution and genetic variation.

10.5.3 Performance Metrics

We investigate the performance metrics such as playback quality and playback quality fluctuation represented by PSNR and standard deviation of PSNR, and average energy consumption of our proposed RDO solution. In addition, we compare the RDO solution with the other two nonoptimal rate and description schemes, i.e., equal performance (EQU) and random performance (RAN). For EQU, the rate and the number of descriptions are equal to $\frac{R_j^{(M)} \times m}{M}$ and $\frac{\sum_{h=0}^{H_i} d_{i,j}^{(h)}}{H_i + 1}$, respectively, in which $R_j^{(M)}$ and $d_{i,j}^{(h)}$ are the optimal maximum encoding rate and the optimal number of descriptions found by solving (10.19) and (10.20). Meanwhile, for RAN, the rate and the number of descriptions are randomly allocated and distributed in the range of $(0, R_j^{(M)}]$ and $(0, \sum_{h=0}^{H_i} d_{i,j}^{(h)}]$, respectively.

10.5.3.1 Performance Metrics Versus α

To evaluate the performance metrics versus the skewed access rate (α) in the range from 0 to 1, we set $\varepsilon_{i,j}^{(h)} = 0.1$, $E = 10$ Joules, $H = 3$, and without the effect of interference constraint in (10.20). Figure 10.4a shows that our RDO outperforms the EQU and RAN on higher PSNR for better quality playback, i.e., up to 8.5 dB and 8.1 dB higher compared to the EQU and RAN, respectively. RDO's performance gets better with a higher value of α among the segments. The RDO also provides lower playback quality fluctuation for smoother playback than the EQU and RAN, as can be seen in Figure 10.4b. Importantly, Figure 10.4c shows that the RDO strictly follows the energy constraint (i.e., $\overline{E} \leq E = 10$ Joules) to keep the overall average energy consumption of RDO lower for higher energy savings than the EQU and RAN. The PSNR is relatively high because all descriptions are protected by FEC.

10.5.3.2 Performance Metrics Versus $\varepsilon_{i,j}^{h}$

We set $\alpha = 0.75$, $E = 10$ Joules, $H = 3$, and without the effect of interference constraint in (10.20), to evaluate the performance metrics versus $\varepsilon_{i,j}^{(h)}$ in the range from 0.05 to 0.5. In Figure 10.5, our RDO outperforms the EQU and RAN with higher PSNR for better quality playback (Figure 10.5a), lower playback quality fluctuation for smoother playback (Figure 10.5b), and less overall average energy consumption (Figure 10.5c). The playback quality fluctuation of RDO in Figure 10.5b is not always lower than that of EQU and RAN because, for the purpose of mitigating the complexity of GA, we have not taken into account the fluctuation constraint in the RDO problem.

10.5.3.3 Performance Metrics Versus E

We evaluate the performance metrics under the effect of energy constraint E in the range from 2.5 to 20 Joules by keeping $\alpha = 0.75$, $\varepsilon_{i,j}^{(h)} = 0.1$, $H = 3$, and without the effect of interference constraint in (10.20). Figure 10.6 shows that RDO surpasses EQU and RAN, especially when E increases, the VHs and the MBS transmit more

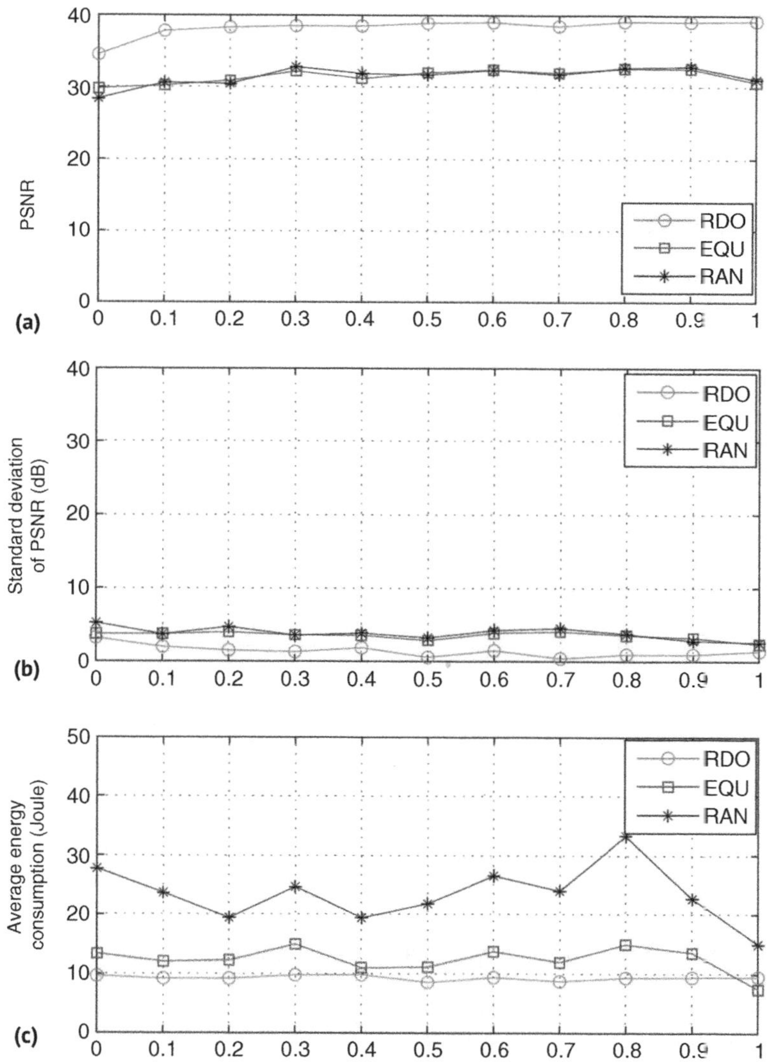

Figure 10.4 Performance versus α: (a) PSNR; (b) standard deviation of PSNR; (c) average energy consumption. *Source:* [1]/IEEE/CC BY 4.0.

descriptions to the RU, thereby providing higher PSNR performance. Thus, our proposed solution can adaptively change the energy constraint depending on the energy condition of VHs and MBS to serve high playback quality.

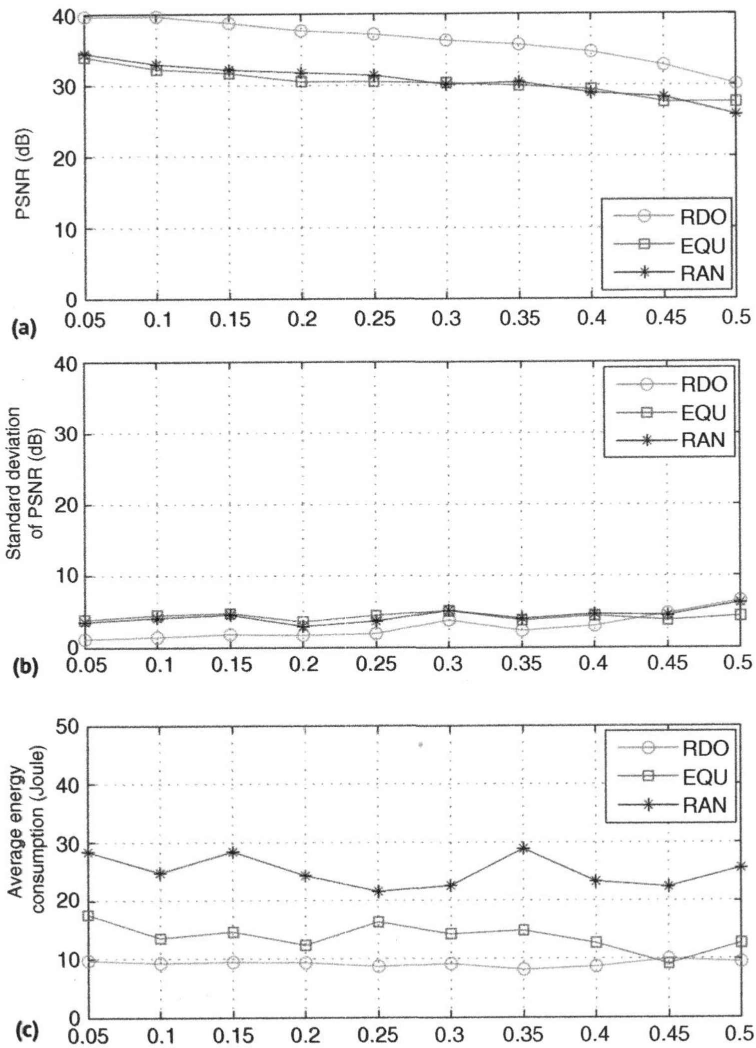

Figure 10.5 Performance versus $\varepsilon_{i,j}^{(h)}$: (a) PSNR; (b) standard deviation of PSNR; (c) average energy consumption. *Source:* [1]/IEEE/CC BY 4.0.

10.5.3.4 Performance Metrics Versus $\gamma^{(0)}$ and H

To investigate how the target SINR ($\gamma^{(0)}$) and the number of GoHs (H) impact the system performance, we consider the interference constraint in (10.20); change $\gamma^{(0)}$ from 0 dB to 30 dB and H from 1 to 5; and set $\alpha = 0.75$, $E = 10$ Joules, and $\varepsilon_{i,j}^{(h)} = 0.1$. In Figure 10.7a, if $\gamma^{(0)}$ increases, i.e., the effect of interference of GoHs

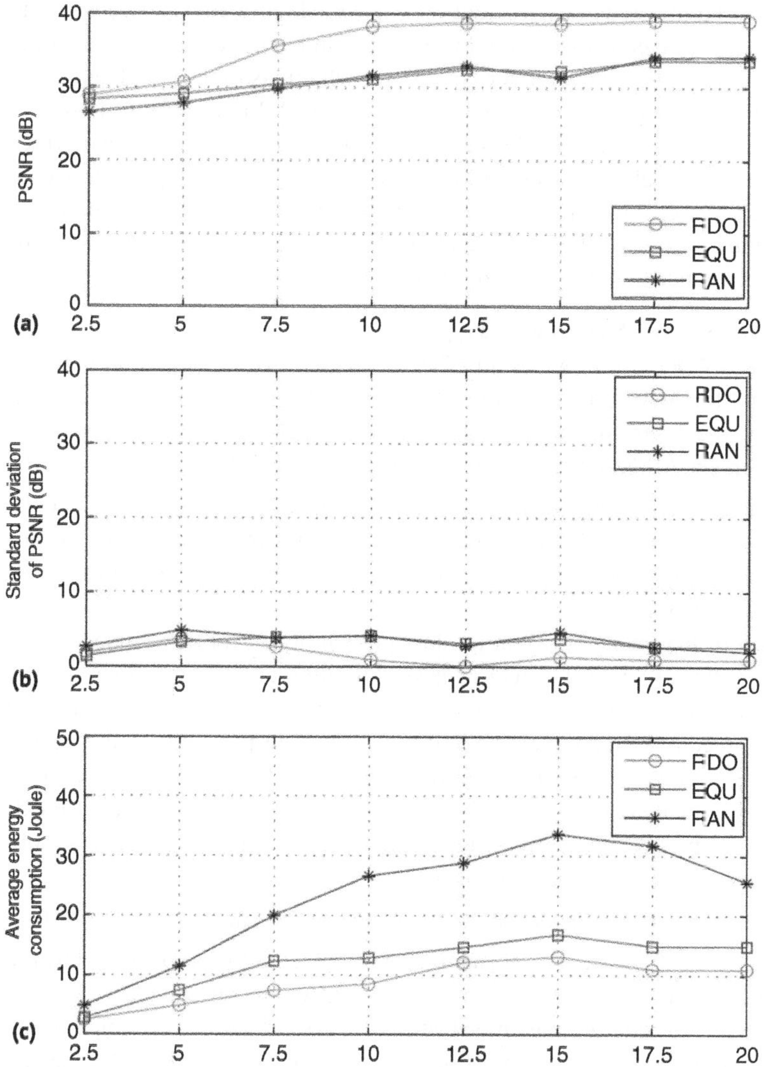

Figure 10.6 Performance versus E: (a) PSNR; (b) standard deviation of PSNR; (c) average energy consumption. *Source:* [1]/IEEE/CC BY 4.0.

on the SHs is more strictly limited, the common transmission powers of all GoHs decrease proportionally to the optimal transmission rate $R_j^{(M)}$ given in (10.15). The quality of received video, therefore, gets lower as $\gamma^{(0)}$ increases. Meanwhile, the increase of $\gamma^{(0)}$ does not have a significant effect on the quality fluctuation (Figure 10.7b) and energy consumption (Figure 10.7c). Importantly, when the density of VHs is higher, increasing H, the quality of received video is higher (Figure 10.7a) and the quality fluctuation is lower (Figure 10.7b) because more opportunities are given to distribute the optimal number of descriptions to proper GoHs that have better characteristics of wireless channels to the RU. The quality of received video gets saturated if we keep increasing H. In dense D2D networks, the number of GoHs is strictly selected to satisfy both the quality of received video and the complexity of GA as shown in Algorithm 10.1. The average energy consumption decreases with the increasing number of GoHs since the total energy consumption is shared out between them. In comparison, the proposed RDO gains higher performance than the EQU and RAN.

10.5.3.5 Performance Metrics Versus $\varepsilon_{i,j}^{(0)}$

The proposed RDO also provides an efficient cooperative transmission between the VHs and the MBS. To demonstrate how the VHs and the MBS cooperate in transmitting the descriptions, we change the lossy characteristic of wireless channels between the MBS and the RU $\varepsilon_{i,j}^{(0)}$ from 0.05 to 1, while keeping $H = 3$, the lossy characteristic of wireless channels between the VHs and the RU $\varepsilon_{i,j}^{(h)} = 0.75$, $h = 1, 2, 3$, and $\alpha = 0.75$. To fully observe this cooperative transmission, we further relax the energy constraint by setting $E = 100$ Joules and do not consider the interference constraint in (10.20).

As shown in Figure 10.8a, when the loss probability $\varepsilon_{i,j}^{(0)}$ of the wireless channel between the MBS and the RU increases, the number of descriptions sent by the MBS decreases, whereas that sent by the VHs increases, in order to minimize the average reconstructed distortion and maintain high playback quality. Obviously, the PSNR (Figure 10.8b) decreases and the quality fluctuation (Figure 10.8c) increases with the increase of $\varepsilon_{i,j}^{(0)}$. In comparison, our RDO achieves higher PSNR than the EQU and RAN but suffers higher quality fluctuation when the loss probability of description is too high, i.e., $\varepsilon_{i,j}^{(h)} > 0.5$.

In Figure 10.8d, the proposed RDO consumes more energy when $\varepsilon_{i,j}^{(0)} < 0.75$ because if $\varepsilon_{i,j}^{(0)} < \varepsilon_{i,j}^{(h)}$, $h = 1, 2, \ldots, H_i$, a larger portion of energy is consumed by the MBS rather than by the VHs to minimize the average reconstructed distortion. When $\varepsilon_{i,j}^{(0)} \geq \varepsilon_{i,j}^{(h)}$, the energy consumed by the VHs to send one description is much less than that by the MBS, most of the descriptions are sent by the VHs, causing RDO to consume less energy than EQU and RAN.

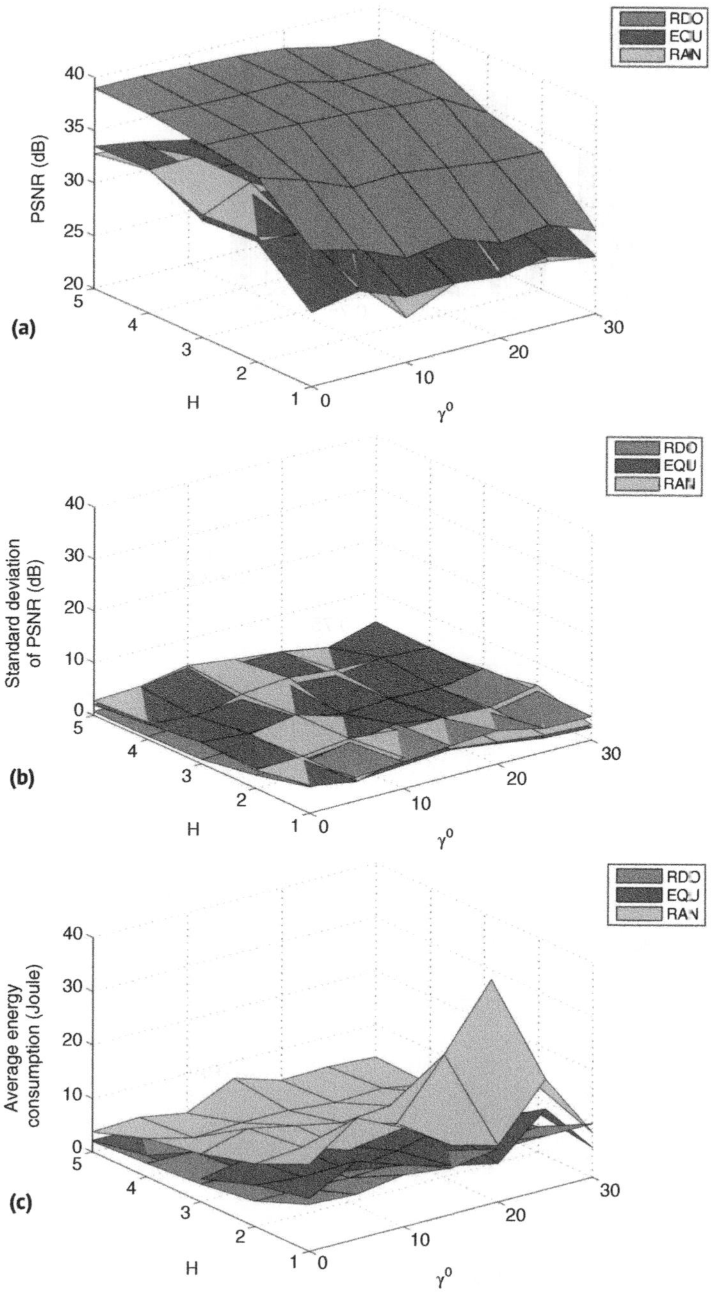

Figure 10.7 Performance versus $\gamma^{(0)}$ and H: (a) PSNR; (b) standard deviation of PSNR; (c) average energy consumption. *Source:* [1]/IEEE/CC BY 4.0.

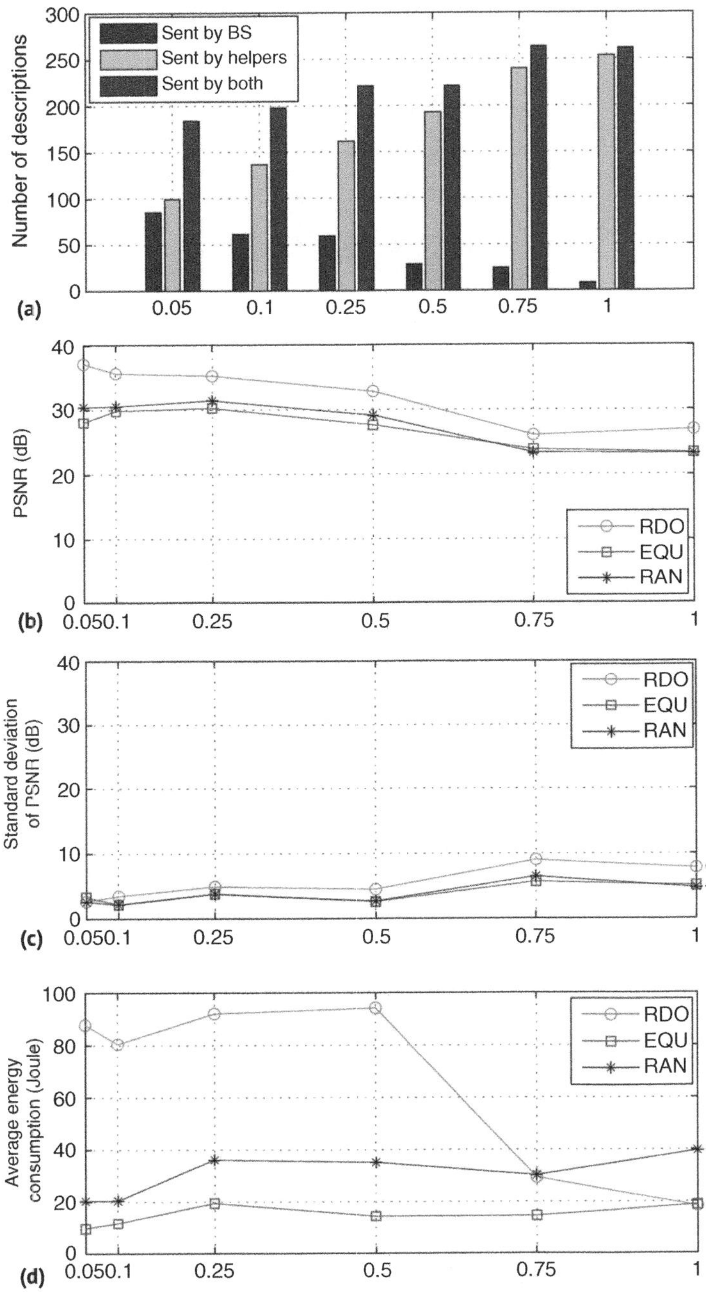

Figure 10.8 Performance versus $\varepsilon_{i,j}^{(0)}$: (a) number of descriptions; (b) PSNR; (c) standard deviation of PSNR; (d) average energy consumption. *Source:* [1]/IEEE/CC BY 4.0.

10.6 Conclusion

In this chapter, we have proposed a joint RDO solution for CVS in dense D2D networks. Particularly, the VHs in different GoHs cooperated with the MBS to provide the RUs with high QoE of video streaming in two modes: D2D underlay and normal cellular communications. High QoE was represented by a high PSNR for playback quality and a low quality fluctuation for smooth playback. Furthermore, the proposed solution consumed less energy and preserves the interference effect of D2D communications on the SHs. The idea behind the benefits of the proposed solution is that we take into account many aspects of lossy wireless channels, intra-popularity of different segments and RDM, LMDC-FEC, and constraints on the remaining energy of the VHs and target SINR of SHs, for high performance of video streaming.

References

1 N.-S. Vo, T. Q. Duong, T. D. Hoang, and A. Kortun. Optimal video streaming in dense 5G networks with D2D communications. *IEEE Access*, 6:209–223. 2017.

2 X. Du, N.-S. Vo, T. Q. Duong, and L. Shu. Joint replication density and rate allocation optimization for VoD systems over wireless mesh networks. *IEEE Trans. Circuits Syst. Video Technol.*, 23(7):1260–1273. 2013.

3 N.-S. Vo, T. Q. Duong, and M. Guizani. QoE-oriented resource efficiency for 5G two-tier cellular networks: A femtocaching framework. In *Proc. IEEE Global Commun. Conf.*, pages 1–6, Washington, DC, Dec. 2016.

4 S.-M. Tseng, S.-T. Wen, C. Fang, and M. Norouzi. Cross layer power allocation by graph neural networks in heterogeneous D2D video communications. *IEEE Access*, 13:44484–44496. 2025.

5 Q.-N. Tran, N.-S. Vo, T.-M. Phan, T. C. Lam, and A. Masaracchia. Multi-rate probabilistic caching optimised video offloading in dense D2D networks. *IEEE Commun. Lett.*, 27(4):1240–1244. 2023.

6 X. Zhang, X. Wei, L. Zhou, and Y. Qian. Social-content-aware scalable video streaming in Internet of video things. *IEEE Internet Things J.*, 9(1):830–843. 2022.

7 S. Chatterjee and S. De. QoE-aware cross-layer adaptation for D2D video communication in cooperative cognitive radio networks. *IEEE Syst. J.*, 16(2):2078–2089. 2022.

8 S. Ullah, K. Kim, A. Manzoor, L. U. Khan, S. M. A. Kazmi, and C. S. Hong. Quality adaptation and resource allocation for scalable video in D2D communication networks. *IEEE Access*, 8:48060–48073. 2020.

9 J. Ma, L. Liu, H. Song, R. Shafin, B. Shang, and P. Fan. Scalable video transmission in cache-aided device-to-device networks. *IEEE Trans. Wireless Commun.*, 19(6):4247–4261. 2020.

10 I. Ahmed, M. H. Ismail, and M. S. Hassan. A QoE-aware mode selection framework for video streaming in D2D networks. *IEEE Access*, 8:169272–169285. 2020.

11 P. A. Chou, H. J. Wang, and V. N. Padmanabhan. Layered multiple description coding. In *Proc. Packet Video Workshop*, pages 1–7, Nantes, France, Apr. 2003.

12 W. Xiang, C. Zhu, C. K. Siew, Y. Xu, and M. Liu. Forward error correction-based 2-D layered multiple description coding for error-resilient H.264 SVC video transmission. *IEEE Trans. Circuits Syst. Video Technol.*, 19(12):1730–1738. 2009.

13 G. J. Sullivan, J.-R. Ohm, W.-J. Han, and T. Wiegand. Overview of the high efficiency video coding (HEVC) standard. *IEEE Trans. Circuits Syst. Video Technol.*, 22(12):1649–1668. 2012.

14 J. Vanne, M. Viitanen, T. D. Hamalainen, and A. Hallapuro. Comparative rate-distortion-complexity analysis of HEVC and AVC video codecs. *IEEE Trans. Circuits Syst. Video Technol.*, 22(12):1885–1898. 2012.

15 J. M. Boyce, J. Chen, Y. Ye, and A. K. Ramasubramonian. Overview of SHVC: Scalable extensions of the high efficiency video coding standard. *IEEE Trans. Circuits Syst. Video Technol.*, 26(1):20–34. 2016.

16 D. Jurca, P. Frossard, and A. Jovanovic. Forward error correction for multipath media streaming. *IEEE Trans. Circuits Syst. Video Technol.*, 19(9):1315–1326. 2009.

17 R. Puri and K. Ramchandran. Multiple description source coding using forward error correction codes. In *Proc. Asilomar Signals, Syst., Comput. Conf.*, pages 342–346, Pacific Grove, CA, Oct. 1999.

18 L. Breslau, P. Cao, L. Fan, G. Phillips, and S. Shenker. Web caching and Zipf-like distributions: Evidence and implications. In *Proc. IEEE Int. Conf. on Computer Commun.*, pages 126–134, New York, NY, Mar. 1999.

19 S. Shalmashi, G. Miao, and S. B. Slimane. Interference management for multiple device-to-device communications underlaying cellular networks. In *Proc. IEEE Personal, Indoor and Mobile Radio Commun.*, pages 223–227, London, United Kingdom, Sep. 2013.

20 L. B. Le. Fair resource allocation for device-to-device communications in wireless cellular networks. In *Proc. IEEE Global Commun. Conf.*, pages 5451–5456, Anaheim, CA, Dec. 2012.

21 T. Fang and L. P. Chau. GOP-based channel rate allocation using genetic algorithm for scalable video streaming over error-prone networks. *IEEE Trans. Image Process.*, 15(6):1323–1330. 2006.

22 L. Booker. Improving Search in Genetic Algorithms. In *Genetic Algorithms and Simulated Annealing*, L. Davis (Ed.), pp. 61–73, Morgan Kaufmann Publishers, 1987.

23 D. E. Goldberg. *Genetic Algorithms in Search, Optimization, and Machine Learning*. Addison-Wesley, Reading, MA, USA, 1988.

24 J. Wu, B. Cheng, M. Wang, and J. Chen. Energy-aware concurrent multipath transfer for real-time video streaming over heterogeneous wireless networks. *IEEE Trans. Circuits Syst. Video Technol.*, 28(8):2007–2023. 2018.

25 C. Bezerra, A. D. Carvalho, D. Borges, N. Barbosa, J. Pontes, and E. Tavares. QoE and energy consumption evaluation of adaptive video streaming on mobile device. In *Proc. IEEE Annu. Consum. Commun. Netw. Conf.*, pages 1–6, Las Vegas, NV, Jan. 2017.

26 N. Eshraghi, V. Shah-Mansouri, and B. Maham. QoE-aware power allocation for device-to-device video transmissions. In *Proc. IEEE Personal, Indoor and Mobile Radio Commun.*, pages 1–5, Valencia, Spain, Sep. 2016.

11

Cross-layer Optimization Design for Multimedia Applications and Services in Content Delivery Networks

11.1 Introduction

The rapid growth of Internet of Things (IoT) connections—ranging from smart transportation and healthcare monitoring to smart cities,[1] especially in immersive MAS such as extended reality (encompassing virtual, augmented, and mixed reality)—has driven an explosive surge in mobile data traffic and imposed stringent demands on system performance [2, 3]. Various MASs with diverse QoS and QoE requirements at high data rate and low latency pose significant challenges for CDNs. In the era of 6G, it is more challenging for CDN due to varying wireless environments and the diverse devices and behaviors of dense mobile users (MUs).

While conventional single-layer design (SLD) in wireless networks typically optimizes each protocol layer separately, often leading to suboptimal solutions for the varying, complicated, and heterogeneous requirements of 6G networks, CLD offers a more robust and effective approach [4]. CLD enables a joint optimization across multiple layers, e.g., application (APP), data link (DLK), network (NET), and physical (PHY) layers. This approach provides greater flexibility to address the stringent and diverse QoS and QoE requirements, while efficiently utilizing the scarce resources (e.g., spectrum and energy) [3, 5, 6] and adapting well to the highly dynamic wireless channels and characteristics of multimedia contents [4, 6, 7]. Consequently, CLD not only outperforms SLD in terms of QoS and QoE, but also facilitates scalable and intelligent system management for MAS in 6G CDN.

Recently, CLD for MAS in 6G networks has attracted considerable research attention thanks to its potential to meet the stringent and diverse QoS and QoE requirements. Particularly, the work in [8] introduces a CLD for scalable video transmission in the Internet of vehicle scenarios over 6G networks. The authors

---

1 This chapter has been published partly in [1].

utilized the modulation at the PHY layer, power allocation at the DLK layer, and video encoding structure with bitrate control at the APP layer to achieve high playback quality of the received videos. In [9], an edge-selective sharing architecture for massive mobile video streaming is proposed by developing a CLD framework. The framework integrates user scheduling and power allocation at the DLK layer, along with buffering at the NET layer, to reduce the probability of rebuffering events, and thus enhance the video streaming performance. Considering device-to-device communications, a joint scheme of source encoder rate adaptation at the APP layer with beamforming and power control at the PHY layer to enhance the playback quality of the received videos was proposed in [3].

To achieve high offloading performance, a CLD is deployed for multiple-input single-output video communication systems assisted by reconfigurable intelligent surfaces [6]. It considers video parameters at the APP layer, resource allocation at the DLK layer, and channel state information at the PHY layer. The challenges in real-time extended reality video transmission are addressed by a QoE-oriented CLD framework studied in [2]. Adaptive bitrate at the APP layer and wireless resource scheduling at the DLK layer were deployed to maximize the QoE. Importantly, a CLD is implemented in the form of a management framework to enhance the extended reality system security in zero-trust wireless communications [10]. The study combines adaptive resource allocation and diverse transmission routes at the NET layer and DLK layer with physical layer security to secure the MUs while ensuring high QoE.

However, most studies on CLD have not addressed the dynamic levels of QoS and QoE across different classes of MAS (i.e., real-time and non-real-time traffic), nor have they provided specific case studies to demonstrate the role of CLD [2, 3, 6, 8–10]. In this chapter, bursty characteristics of different multimedia traffic classes at the APP layer, queuing mechanisms at the DLK layer, and AMC at the PHY layer are jointly optimized in a Markov-based CLD framework. The framework enables us to investigate key aspects of QoS such as PDR, PER, delay, and throughput. Two specific CLD models are formulated: one for general MAS and another for real-time video streaming, aiming to minimize PDR and reconstructed distortion while ensuring smooth playback for high QoS and QoE. Simulation results demonstrate the benefits of the proposed CLD and provide useful guidelines for CLD of MAS in 6G CDN.

The rest of this chapter is organized as follows. In Section 11.2, we introduce the detailed CLD framework for MAS. The analysis and formulation of CLD are presented in Section 11.3. Section 11.4 formulates the CLD optimization problem and its solution. The performance evaluations are conducted in Section 11.5. Finally, Section 11.6 concludes the chapter.

11.2 CLD Framework for MAS

The CLD integrates the APP, DLK, and PHY layers of the protocol stack. In this CLD, time-division multiple access is applied for downlink communications between the base stations (BSs) and MUs over time-varying Nakagami-m fading channels. Assuming that each device is equipped with a single antenna and a channel is modelled as quasi-static fading, i.e., it remains unchanged within a frame and varies independently across the frames. We further adopt a discrete-time model with an infinite sequence of equal periods, each corresponding to a frame duration T_f. The key notations are summarized in Table 11.1.

At the APP layer, we consider multimedia communications in which the MUs requesting different traffic classes can be simultaneously served by the BS. The APP layer supports MAS with multiple traffic classes, categorized into real-time and non-real-time classes. The real-time class imposes stricter delay requirements than the non-real-time class, giving real-time packets higher transmission priority. To model the multimedia traffic, we adopt the Poisson process and Markov-modulated Poisson process (MMPP) [11]. At the DLK layer, we propose a buffer partitioning mechanism to handle the two traffic classes. Finally, at the PHY layer,

Table 11.1 Notations.

Symbols	Descriptions
$\mathbf{P}^s$	Transition matrix of the whole system
T_f	Frame duration in a discrete time model
λ_{A}	Average arrival rate
q_{A}	$\mathsf{A} \in \{\mathsf{H}, \mathsf{L}\}$, average number of packets of high/low priority traffic arriving within T_f
$\mathbf{P}^{tr}$	$N_{\mathsf{H}} \times N_{\mathsf{H}}$ state transition matrix
$\mathrm{P}_{ar}(q_{\mathsf{A}}\|\lambda_{\mathsf{A}})$	Arrival probability of q_{A} packets without blocking
$\mathrm{P}_{bl}(q_{\mathsf{A}}\|\lambda_{\mathsf{A}})$	Blocking probability of newly incoming packets
B	Total buffer size
N_{AMC}	Number of AMC modes
PER_n	Approximation of packet error rate of ACM mode n, $n = 1, 2, \ldots, N_{\mathrm{AMC}}$
P^{ch}	State transition matrix of PHY layer
π^{s}	Steady state probability vector of system
$D_d^{\mathsf{A}}, D_e^{\mathsf{A}}, \eta^{\mathsf{A}}$	Packet drop rate, packet error rate, and packet loss rate
D^{A}	Total delay
$\overline{d}$	Average reconstructed distortion of received video

AMC is employed with N_{AMC} modes to support multiple transmission rates. Each layer is described in detail as follows.

11.2.1 APP Layer

At the APP layer, the low priority of non-real-time traffic (class L) is modelled as a Poisson process with arrival rate λ_{L}, whereas the high priority of real-time traffic (class H) is modelled as a multiple state MMPP. In this way, the real-time traffic is described by a discrete Markov chain with N_{H} states. An illustrative example of a two-state MMPP is given by

$$\mathbf{P}^{tr} = \begin{bmatrix} \mathsf{P}^{tr}_{1,1} & \mathsf{P}^{tr}_{1,2} \\ \mathsf{P}^{tr}_{2,1} & \mathsf{P}^{tr}_{2,2} \end{bmatrix}, \tag{11.1}$$

where $\mathbf{P}^{tr}$ is the $N_{\mathsf{H}} \times N_{\mathsf{H}}$ state transition matrix and element (i, j), denoted by $P^{tr}_{i,j}$, represents the transition probability from state i with arrival rate $\lambda_{\mathsf{H},i}$ to state j with arrival rate $\lambda_{\mathsf{H},j}$.

We further denote q_{A}, $\mathsf{A} \in \{\mathsf{H}, \mathsf{L}\}$, as the number of packets of high/low priority traffic arriving in a discrete interval T_f. Assume that q_{A} follows a Poisson distribution with parameter $\lambda_{\mathsf{A}} T_f$, i.e., $\lambda_{\mathsf{A}} = \lambda_{\mathsf{L}}$ for non-real-time traffic and $\lambda_{\mathsf{A}} = \lambda_{\mathsf{H},i}$ for real-time traffic, the probability that q_{A} packets arrive during a frame duration T_f is written as

$$\mathsf{P}_{ar}\left(q_{\mathsf{A}} | \lambda_{\mathsf{A}}\right) = \frac{\left(\lambda_{\mathsf{A}} T_f\right)^{q_{\mathsf{A}}}}{q_{\mathsf{A}}!} \exp\left(-\lambda_{\mathsf{A}} T_f\right). \tag{11.2}$$

Due to the finite buffer length, an overflow situation may occur at the lower DLK layer, and thus, when the buffer is full, newly arriving packets are discarded. To illustrate this overflow situation, $\mathsf{P}_{bl}\left(q_{\mathsf{A}} | \lambda_{\mathsf{A}}\right)$ is defined as the probability that q_{A} packets may arrive to fulfil the available buffer space within T_f. So, any additional incoming packets, e.g., the $(q_{\mathsf{A}} + 1)$-th, $(q_{\mathsf{A}} + 2)$-th, ... , are discarded from the buffer once it is full. Using the law of total probability, the blocking probability $\mathsf{P}_{bl}\left(q_{\mathsf{A}} | \lambda_{\mathsf{A}}\right)$ is given by

$$\mathsf{P}_{bl}\left(q_{\mathsf{A}} | \lambda_{\mathsf{A}}\right) = \begin{cases} 1 - \sum_{t=0}^{q_{\mathsf{A}}} \mathsf{P}_{ar}\left(t | \lambda_{\mathsf{A}}\right) & \text{if} \quad q_{\mathsf{A}} \geq 0, \\ 1 & \text{if} \quad q_{\mathsf{A}} < 0. \end{cases} \tag{11.3}$$

11.2.2 DLK Layer

At the DLK layer, the data is processed packet by packet. A buffer with first-in first-out scheduling is implemented at the BS. The packets arriving from the APP layer are stored in a finite-length buffer of total size B. Assuming an integration of L and H priority traffic, the non-real-time and real-time classes are assigned the buffer sizes of B_L and B_H, respectively (where $B = B_H + B_L$). To guarantee the QoS requirements of both non-real-time and real-time traffic classes, B_H and B_L are optimally allocated according to their arrival rates λ_A, such that the sum of their blocking probabilities in (11.3) is minimized, thereby reducing the total PDR of the system.

In our buffer scheduling model, real-time traffic, which requires lower delay, is given higher priority for transmission over the PHY channel than non-real-time traffic. Hence, a strictly preemptive priority mechanism is deployed between the two classes. This mechanism leads to two cases, depending on the PHY channel condition and the buffer-H state. The buffer-L is served only when no high-priority packets are waiting in the buffer-H and the channel is available, i.e., the channel state $n \neq 0$. This approach may lead to a suboptimal solution, where low-priority traffic is not served even though the channel is still capable of delivering additional packets.

It is important to efficiently utilize network resources. One possible solution is to apply a buffer scheduling mechanism that mixes both low- and high-priority traffic classes under given channel conditions. When the number of high-priority packets is smaller than the channel capacity, the remaining slots can be used to transmit both classes. This strategy is detailed in Section 11.3 through the construction of the Markov chain transition matrix.

11.2.3 PHY Layer

At the PHY layer, data is processed in frames, which consist of multiple packets from the DLK layer. To exploit the time-varying nature of the PHY channel, assuming AMC with adaptive modulation and convolutional coding is adopted [12, Table 2]. As a result, depending on the specific transmission mode, a frame may consist of one or more packets. Given a rate R_n (bits/symbol) associated with transmission mode n, $n = 1, 2, \dots, N_{AMC}$, we can obtain the number of packets per frame K_n as computed in [13]. This assumption, which simplifies our analytical framework, has been widely adopted in the literature [12, 14, 15]. Let us denote the instantaneous received signal-to-noise ratio (SNR) as γ, since we investigate for a Nakagami-m fading channel with fading severity parameter $m \in [1/2, \infty)$, the probability density function (PDF) of γ is expressed as

$$f_\gamma(\gamma) = \frac{\gamma^{m-1}}{\Omega^m \Gamma(m)} \exp\left(-\frac{\gamma}{\Omega}\right), \tag{11.4}$$

where $\Omega = \gamma_0/m$, γ_0 is the average SNR, and $\Gamma(\cdot)$ is the Gamma function.

The objective of AMC is to maximize spectral efficiency while ensuring that the PER remains below a given threshold P_0. To do so, mode n employs convolutional code n combined with M_n-QAM modulation at a transmission rate of R_n. Hence, the SNR range is divided into $N_{\mathsf{AMC}}+1$ non-overlapping intervals, each defined by two thresholds, γ_n and γ_{n+1}. Mode n is employed when the instantaneous SNR γ falls within $[\gamma_n, \gamma_{n+1})$. Note that mode 0, i.e., $\gamma_0 \le \gamma < \gamma_1$, corresponds to a system outage without payload data being transmitted. To compute the thresholds, we adopt a similar approach to [13] by approximating PER for mode n under additive white Gaussian noise as

$$\mathrm{PER}_n = \begin{cases} 1 & \text{if } 0 < \gamma < \tilde{\gamma}_n, \\ a_n \exp(-g_n\gamma) & \text{if } \gamma \ge \tilde{\gamma}_n, \end{cases} \tag{11.5}$$

where a_n, g_n, and $\tilde{\gamma}_n$ are computed by least-square fitting to the exact PER curve [12, Table 2]. The average PER of mode n over Nakagami-m fading channels, $\overline{\mathrm{PER}}_n$, can be obtained by integrating (11.5) over the PDF of γ.

So far, the thresholds $\gamma_{n\,n=0}^{\,N_{\mathsf{AMC}}+1}$ can be determined to satisfy $\overline{\mathrm{PER}}_n \le P_0$ using the following search algorithm:

- Step 1: Set $n = N$ and $\gamma_{N_{\mathsf{AMC}}+1} = \infty$.
- Step 2: Search for γ_n within $[0, \gamma_{n+1})$ such that $\overline{\mathrm{PER}}_n = P_0$.
- Step 3: If $n > 1$, then set $n = n - 1$ and go back to Step 2; otherwise, go to Step 4.
- Step 4: Set $\gamma_0 = 0$.

Next, we model the channel using a finite-state Markov chain [16], with $N_{\mathsf{AMC}}+1$ states. Let P^{ch}, of size $(N_{\mathsf{AMC}}+1)\times(N_{\mathsf{AMC}}+1)$, denote the PHY-layer state transition matrix encompassing both the radio channel and the AMC scheme. Under the slow-fading assumption, transitions occur only between adjacent states, that is $\mathsf{P}^{ch}_{n,l} = 0$, $\forall |l-n| > 1$. And thus, the transition probability between two adjacent states is given by

$$\begin{aligned} \mathsf{P}^{ch}_{n,n+1} &= \frac{\mathsf{LCR}_{n+1}T_f}{\mathrm{Pr}_n}, \qquad n = 0, \dots, N_{\mathsf{AMC}} - 1, \\ \mathsf{P}^{ch}_{n,n-1} &= \frac{\mathsf{LCR}_n T_f}{\mathrm{Pr}_n}, \qquad n = 1, \dots, N_{\mathsf{AMC}}, \end{aligned} \tag{11.6}$$

where Pr_n is the probability that mode n is chosen, given by $\mathrm{Pr}_n = \int_{\gamma_n}^{\gamma_{n+1}} f_\gamma(\gamma)\, d\gamma$. Moreover, let LCR_n denote the level crossing rate of mode n, expressed as

$$\mathsf{LCR}_n = \sqrt{\frac{2\pi\gamma_n}{\Omega}}\,\frac{f_d}{\Gamma(m)}\left(\frac{\gamma_n}{\Omega}\right)^{m-1} \exp\left(-\frac{\gamma_n}{\Omega}\right), \tag{11.7}$$

where f_d is the Doppler frequency, the probability for mode n to remain unchanged is given by

$$\mathsf{P}^{ch}_{n,n} = \begin{cases} 1 - \mathsf{P}^{ch}_{n,n+1} - \mathsf{P}^{ch}_{n,n-1} & \text{if } 0 < n < N_{\mathsf{AMC}} \\ 1 - \mathsf{P}^{ch}_{0,1} & \text{if } n = 0 \\ 1 - \mathsf{P}^{ch}_{N_{\mathsf{AMC}},N_{\mathsf{AMC}}-1} & \text{if } n = N_{\mathsf{AMC}} \end{cases}. \tag{11.8}$$

The average PER at the PHY layer can be minimized to improve system performance by taking (11.8) and $\overline{\text{PER}}_n$ into account. In other words, $\sum_{n=1}^{N_{\mathsf{AMC}}} \mathsf{P}^{ch}_{n,n}\overline{\text{PER}}_n$ can be minimized by optimally assigning $\overline{\text{PER}}_n$, i.e., by setting switching threshold γ_n, according to the corresponding value of $\mathsf{P}^{ch}_{n,n}$.

11.3 Analysis and Formulation of CLD Optimization Framework

In this section, we construct the embedded Markov chain based on the system configuration in Section 11.2. A four-dimensional Markov chain is proposed to model the CLD, which provides sufficient information for system performance evaluation. We divide the time into equal intervals, each of duration T_f. Let $(n, i, q_{\mathsf{H}}, q_{\mathsf{L}})$ denote the state space of the system where $n \in [0, N_{\mathsf{AMC}}]$ is the current channel state, $i \in [1, N_{\mathsf{H}}]$ is the state of real-time traffic, and $q_{\mathsf{H}} \in [0, B_{\mathsf{H}}]$ and $q_{\mathsf{L}} \in [0, B_{\mathsf{L}}]$ are the number of high- and low-priority packets in the buffer-H and buffer-L, respectively.

It can be seen from (11.1) and (11.8), the transition from the state $(n, i, q_{\mathsf{H}}, q_{\mathsf{L}})$ to $(k, j, v_{\mathsf{H}}, v_{\mathsf{L}})$ is Markovian, thus the queuing system can be described by an embedded Markov chain. The objective is to generate the transition matrix, $\mathbf{P}^s$, of the Markov chain for the whole system. To do so, it is necessary to compute all possible transition probabilities. We construct the system transition matrix $\mathbf{P}^s$ by first considering real-time traffic, followed by non-real-time traffic. Specifically, we discuss how to calculate the transition probability $\mathsf{P}^s_{(n,i,q_{\mathsf{H}},q_{\mathsf{L}}),(k,j,v_{\mathsf{H}},v_{\mathsf{L}})}$ from the current state $(n, i, q_{\mathsf{H}}, q_{\mathsf{L}})$ to the next state $(k, j, v_{\mathsf{H}}, v_{\mathsf{L}})$.

11.3.1 Real-time Traffic

For real-time traffic, whenever the channel is available, i.e., $n \neq 0$, and the buffer-H is not empty, the data is ready to be transmitted to the destination. In general, four cases for real-time traffic are considered as follows.

- Case 1: Initially, the buffer-H is empty ($q_{\mathsf{A}} = 0$), and the only process is the arrival of packets from the upper layer. When the buffer size exceeds the number of arriving packets, no overflow occurs. In other words, if the buffer does not overflow, i.e., $\upsilon_{\mathrm{H}} \in [0, B_{\mathrm{H}} - 1]$ and $\upsilon_{\mathrm{L}} \in [0, B_{\mathrm{L}} - 1]$, then

$$\mathsf{P}^{s}_{(n,i,0,0),(k,j,\upsilon_{\mathrm{H}},\upsilon_{\mathrm{L}})} = \mathsf{P}^{ch}_{n,k}\mathsf{P}^{tr}_{i,j}\mathsf{P}_{ar}\left(\upsilon_{\mathrm{H}}|\lambda_{\mathrm{H},i}\right)\mathsf{P}_{ar}\left(\upsilon_{\mathrm{L}}|\lambda_{\mathrm{L}}\right). \tag{11.9}$$

When the buffer-H is full, any additional arriving packets are discarded. The corresponding probability is represented by $\mathsf{P}_{bl}\left(B_{\mathrm{H}} - 1|\lambda_{\mathrm{H},i}\right)$, as given in (11.3). Equivalently, if $\upsilon_{\mathrm{H}} = B_{\mathrm{H}}$ and $\upsilon_{\mathrm{L}} \in [0, B_{\mathrm{L}} - 1]$, then

$$\mathsf{P}^{s}_{(n,i,0,0),(k,j,B_{\mathrm{H}},\upsilon_{\mathrm{L}})} = \mathsf{P}^{ch}_{n,k}\mathsf{P}^{tr}_{i,j}\mathsf{P}_{bl}\left(B_{\mathrm{H}} - 1|\lambda_{\mathrm{H},i}\right) \times \mathsf{P}_{ar}\left(\upsilon_{\mathrm{L}}|\lambda_{\mathrm{L}}\right). \tag{11.10}$$

Similarly, for the buffer-L, if $\upsilon_{\mathrm{H}} \in [0, B_{\mathrm{H}} - 1]$ and $\upsilon_{\mathrm{L}} = B_{\mathrm{L}}$, then

$$\mathsf{P}^{s}_{(n,i,0,0),(k,j,\upsilon_{\mathrm{H}},B_{\mathrm{L}})} = \mathsf{P}^{ch}_{n,k}\mathsf{P}^{tr}_{i,j}\mathsf{P}_{ar}\left(\upsilon_{\mathrm{H}}|\lambda_{\mathrm{H},i}\right) \times \mathsf{P}_{bl}\left(B_{\mathrm{L}} - 1|\lambda_{\mathrm{L}}\right). \tag{11.11}$$

When both buffers are full, additional packets are discarded. If $\upsilon_{\mathrm{H}} = B_{\mathrm{H}}$ and $\upsilon_{\mathrm{L}} = B_{\mathrm{L}}$, then

$$\mathsf{P}^{s}_{(n,i,0,0),(k,j,B_{\mathrm{H}},B_{\mathrm{L}})} = \mathsf{P}^{ch}_{n,k}\mathsf{P}^{tr}_{i,j}\mathsf{P}_{bl}\left(B_{\mathrm{H}} - 1|\lambda_{\mathrm{H},i}\right) \times \mathsf{P}_{bl}\left(B_{\mathrm{L}} - 1|\lambda_{\mathrm{L}}\right). \tag{11.12}$$

- Case 2: The channel is not in outage and the number of packets in the buffer exceeds the channel capacity, i.e., $n \neq 0$ and $q_{\mathrm{H}} \geq K_n$. In this case, the PHY channel is available for packet transmission. Since the number of packets in the buffer-H exceeds the channel capacity, only high-priority data is served. Thus, the number of packets in the buffer-H becomes $q_{\mathrm{H}} - K_n$. If the incoming packets for both classes do not exceed their buffer capacities, i.e., $\upsilon_{\mathrm{H}} \in [q_{\mathrm{H}} - K_n, B_{\mathrm{H}} - 1]$ and $\upsilon_{\mathrm{L}} \in [q_{\mathrm{L}}, B_{\mathrm{L}} - 1]$, then

$$\begin{aligned}\mathsf{P}^{s}_{(n,i,q_{\mathrm{H}},q_{\mathrm{L}}),(k,j,\upsilon_{\mathrm{H}},\upsilon_{\mathrm{L}})} &= \mathsf{P}^{ch}_{n,k}\mathsf{P}^{tr}_{i,j} \\ &\times \mathsf{P}_{ar}\left(\upsilon_{\mathrm{H}} - q_{\mathrm{H}} + K_n|\lambda_{\mathrm{H},i}\right)\mathsf{P}_{ar}\left(\upsilon_{\mathrm{L}} - q_{\mathrm{L}}|\lambda_{\mathrm{L}}\right).\end{aligned} \tag{11.13}$$

The three remaining scenarios in Case 2 can be interpreted in the same way as in Case 1: If $\upsilon_{\mathrm{H}} = B_{\mathrm{H}}$ and $\upsilon_{\mathrm{L}} \in [q_{\mathrm{L}}, B_{\mathrm{L}} - 1]$, then

$$\begin{aligned}\mathsf{P}^{s}_{(n,i,q_{\mathrm{H}},q_{\mathrm{L}}),(k,j,B_{\mathrm{H}},\upsilon_{\mathrm{L}})} &= \mathsf{P}^{ch}_{n,k}\mathsf{P}^{tr}_{i,j} \\ &\times \mathsf{P}_{bl}\left(B_{\mathrm{H}} - 1 - q_{\mathrm{H}} + K_n|\lambda_{\mathrm{H},i}\right)\mathsf{P}_{ar}\left(\upsilon_{\mathrm{L}} - q_{\mathrm{L}}|\lambda_{\mathrm{L}}\right).\end{aligned} \tag{11.14}$$

If $\upsilon_{\mathrm{H}} \in [q_{\mathrm{H}} - K_n, B_{\mathrm{H}} - 1]$ and $\upsilon_{\mathrm{L}} = B_{\mathrm{L}}$, then

$$\begin{aligned}\mathsf{P}^{s}_{(n,i,q_{\mathrm{H}},q_{\mathrm{L}}),(k,j,\upsilon_{\mathrm{H}},B_{\mathrm{L}})} &= \mathsf{P}^{ch}_{n,k}\mathsf{P}^{tr}_{i,j} \\ &\times \mathsf{P}_{ar}\left(\upsilon_{\mathrm{H}} - q_{\mathrm{H}} + K_n|\lambda_{\mathrm{H},i}\right)\mathsf{P}_{bl}\left(B_{\mathrm{L}} - 1 - q_{\mathrm{L}}|\lambda_{\mathrm{L}}\right).\end{aligned} \tag{11.15}$$

If $\upsilon_{\mathrm{H}} = B_{\mathrm{H}}$ and $\upsilon_{\mathrm{L}} = B_{\mathrm{L}}$, then

$$\mathsf{P}^{s}_{(n,i,q_{\mathrm{H}},q_{\mathrm{L}}),(k,j,B_{\mathrm{H}},B_{\mathrm{L}})} = \mathsf{P}^{ch}_{n,k}\mathsf{P}^{tr}_{i,j} \times \mathsf{P}_{bl}\left(B_{\mathrm{H}} - 1 - q_{\mathrm{H}} + K_n|\lambda_{\mathrm{H},i}\right)\mathsf{P}_{bl}\left(B_{\mathrm{L}} - 1 - q_{\mathrm{L}}|\lambda_{\mathrm{L}}\right). \tag{11.16}$$

- Case 3: This occurs when $n \neq 0$ and $q_{\mathrm{H}} \in [1, K_n - 1]$. Unlike Case 2, the number of high-priority packets is smaller than the maximum supported by the current transmission mode at the PHY channel. Therefore, some free slots can be exploited for transmitting low-priority data, provided that the number of such packets does not exceed $K_n - q_{\mathrm{H}}$. If the buffer-L size is larger than the capacity permitted by the channel, i.e , $q_{\mathrm{L}} \geq K_n - q_{\mathrm{H}}$, then the number of packets in the buffer-L after transmission must satisfy $\upsilon_{\mathrm{L}} \geq q_{\mathrm{L}} + q_{\mathrm{H}} - K_n$. The same reasoning applies to the reverse scenario: if $q_{\mathrm{L}} < K_n - q_{\mathrm{H}}$, then υ_{L} can take any integer value smaller than B_{L}. Consequently, the condition $\max(0, q_{\mathrm{H}} + q_{\mathrm{L}} - K_n) \leq \upsilon_{\mathrm{L}} \leq B_{\mathrm{L}}$ holds. The corresponding state transition probabilities are obtained similarly to the previous cases and summarized as follows.

If $\upsilon_{\mathrm{H}} \in [0, B_{\mathrm{H}} - 1]$ and $\upsilon_{\mathrm{L}} \in [\max(0, q_{\mathrm{H}} + q_{\mathrm{L}} - K_n), B_{\mathrm{L}} - 1]$ then

$$\mathsf{P}^{s}_{(n,i,q_{\mathrm{H}},q_{\mathrm{L}}),(k,j,\upsilon_{\mathrm{H}},\upsilon_{\mathrm{L}})} = \mathsf{P}^{ch}_{n,k}\mathsf{P}^{tr}_{i,j}\mathsf{P}_{ar}\left(\upsilon_{\mathrm{H}}|\lambda_{\mathrm{H},i}\right) \times \mathsf{P}_{ar}\left(\upsilon_{\mathrm{L}} - \max(0, q_{\mathrm{H}} + q_{\mathrm{L}} - K_n)|\lambda_{\mathrm{L}}\right). \tag{11.17}$$

If $\upsilon_{\mathrm{H}} = q_{\mathrm{H}}$ and $\upsilon_{\mathrm{L}} \in [\max(0, q_{\mathrm{H}} + q_{\mathrm{L}} - K_n), B_{\mathrm{L}} - 1]$, then

$$\mathsf{P}^{s}_{(n,i,q_{\mathrm{H}},q_{\mathrm{L}}),(k,j,B_{\mathrm{H}},\upsilon_{\mathrm{L}})} = \mathsf{P}^{ch}_{n,k}\mathsf{P}^{tr}_{i,j}\mathsf{P}_{bl}\left(B_{\mathrm{H}} - 1|\lambda_{\mathrm{H},i}\right) \times \mathsf{P}_{ar}\left(\upsilon_{\mathrm{L}} - \max(0, q_{\mathrm{H}} + q_{\mathrm{L}} - K_n)|\lambda_{\mathrm{L}}\right). \tag{11.18}$$

If $\upsilon_{\mathrm{H}} \in [0, B_{\mathrm{H}} - 1]$ and $\upsilon_{\mathrm{L}} = B_{\mathrm{L}}$, then

$$\mathsf{P}^{s}_{(n,i,q_{\mathrm{H}},q_{\mathrm{L}}),(k,j,\upsilon_{\mathrm{H}},B_{\mathrm{L}})} = \mathsf{P}^{ch}_{n,k}\mathsf{P}^{tr}_{i,j}\mathsf{P}_{ar}\left(\upsilon_{\mathrm{H}}|\lambda_{\mathrm{H},i}\right) \times \mathsf{P}_{bl}\left(B_{\mathrm{L}} - \max(0, q_{\mathrm{H}} + q_{\mathrm{L}} - K_n)|\lambda_{\mathrm{L}}\right). \tag{11.19}$$

If $\upsilon_{\mathrm{H}} = B_{\mathrm{H}}$ and $\upsilon_{\mathrm{L}} = B_{\mathrm{L}}$, then

$$\mathsf{P}^{s}_{(n,i,q_{\mathrm{H}},q_{\mathrm{L}}),(k,j,B_{\mathrm{H}},B_{\mathrm{L}})} = \mathsf{P}^{ch}_{n,k}\mathsf{P}^{tr}_{i,j}\mathsf{P}_{bl}\left(B_{\mathrm{H}} - 1|\lambda_{\mathrm{H},i}\right) \times \mathsf{P}_{bl}\left(B_{\mathrm{L}} - 1 - \max(0, q_{\mathrm{H}} + q_{\mathrm{L}} - K_n)|\lambda_{\mathrm{L}}\right). \tag{11.20}$$

- Case 4: This case arises when $n = 0$. In other words, the PHY channel is in the outage state, and thus no data transmission is possible. The only event in the system is packet arrival. The state transition probabilities are computed in the same manner as in Case 1, except that the buffer may not be empty, as shown in the following.

If $\upsilon_H \in [q_H, B_H - 1]$ and $\upsilon_L \in [q_L, B_L - 1]$, then

$$\begin{aligned} \mathsf{P}^{s}_{(0,i,q_H,q_L),(k,j,\upsilon_H,\upsilon_L)} &= \mathsf{P}^{ch}_{0,k}\mathsf{P}^{tr}_{i,j} \\ &\times \mathsf{P}_{ar}\left(\upsilon_H - q_H|\lambda_{H,i}\right)\mathsf{P}_{ar}\left(\upsilon_L - q_L|\lambda_L\right). \end{aligned} \tag{11.21}$$

If $\upsilon_H = B_H$ and $\upsilon_L \in [q_L, B_L - 1]$, the

$$\begin{aligned} \mathsf{P}^{s}_{(0,i,q_H,q_L),(k,j,B_H,\upsilon_L)} &= \mathsf{P}^{ch}_{0,k}\mathsf{P}^{tr}_{i,j} \\ &\times \mathsf{P}_{bl}\left(B_H - 1 - q_H|\lambda_{H,i}\right)\mathsf{P}_{ar}\left(\upsilon_L - q_L|\lambda_L\right). \end{aligned} \tag{11.22}$$

If $\upsilon_H \in [q_H, B_H - 1]$ and $\upsilon_L = B_L$, then

$$\begin{aligned} \mathsf{P}^{s}_{(0,i,q_H,q_L),(k,j,\upsilon_H,B_L)} &= \mathsf{P}^{ch}_{0,k}\mathsf{P}^{tr}_{i,j} \\ &\times \mathsf{P}_{ar}\left(\upsilon_H - q_H|\lambda_{H,i}\right)\mathsf{P}_{bl}\left(B_L - 1 - q_L|\lambda_L\right). \end{aligned} \tag{11.23}$$

If $\upsilon_H = B_H$ and $\upsilon_L = B_L$, then

$$\begin{aligned} \mathsf{P}^{s}_{(0,i,q_H,q_L),(k,j,B_H,B_L)} &= \mathsf{P}^{ch}_{0,k}\mathsf{P}^{tr}_{i,j} \\ &\times \mathsf{P}_{bl}\left(B_H - 1 - q_H|\lambda_{H,i}\right)\mathsf{P}_{bl}\left(B_L - 1 - q_L|\lambda_L\right). \end{aligned} \tag{11.24}$$

11.3.2 Non-real-time Traffic

For non-real-time traffic, the system operation corresponds to the last case, as presented in the sequel.

- Case 5: This case occurs when $n \neq 0$, $q_H = 0$, and $q_L > 0$. Owing to the preemptive priority mechanism employed in our CLD framework, the PHY channel can be fully utilized by low-priority data whenever the buffer-H is empty, i.e., $q_H = 0$. Similar to the interpretation in Case 3, we obtain the following.

  If $\upsilon_H \in [0, B_H - 1]$ and $\upsilon_L \in [\max(0, q_L - K_n), B_L - 1]$

$$\begin{aligned} \mathsf{P}^{s}_{(n,i,0,q_L),(k,j,\upsilon_H,\upsilon_L)} &= \mathsf{P}^{ch}_{n,k}\mathsf{P}^{tr}_{i,j} \\ &\times \mathsf{P}_{ar}\left(\upsilon_H|\lambda_{H,i}\right)\mathsf{P}_{ar}\left(\upsilon_L - \max(0, q_L - K_n)|\lambda_L\right). \end{aligned} \tag{11.25}$$

  If $\upsilon_H = B_H$ and $\upsilon_L \in [\max(0, q_L - K_n), B_L - 1]$

$$\begin{aligned} \mathsf{P}^{s}_{(n,i,0,q_L),(k,j,B_H,\upsilon_L)} &= \mathsf{P}^{ch}_{n,k}\mathsf{P}^{tr}_{i,j}\mathsf{P}_{bl}\left(B_H - 1|\lambda_{H,i}\right) \\ &\times \mathsf{P}_{ar}\left(\upsilon_L - \max(0, q_L - K_n)|\lambda_L\right). \end{aligned} \tag{11.26}$$

If $v_{\mathrm{H}} \in [0, B_{\mathrm{H}} - 1]$ and $v_{\mathrm{L}} = B_{\mathrm{L}}$

$$\begin{aligned}\mathrm{P}^{s}_{(n,i,0,q_{\mathrm{L}}),(k,j,v_{\mathrm{H}},B_{\mathrm{L}})} &= \mathrm{P}^{ch}_{n,k}\mathrm{P}^{tr}_{i,j}\mathrm{P}_{ar}\left(v_{\mathrm{H}} - 1|\lambda_{\mathrm{H},i}\right) \\ &\times \mathrm{P}_{bl}\left(B_{\mathrm{L}} - 1 - \max(0, q_{\mathrm{L}} - K_n)|\lambda_{\mathrm{L}}\right).\end{aligned} \tag{11.27}$$

If $v_{\mathrm{H}} = B_{\mathrm{H}}$ and $v_{\mathrm{L}} = B_{\mathrm{L}}$

$$\begin{aligned}\mathrm{P}^{s}_{(n,i,0,q_{\mathrm{L}}),(k,j,B_{\mathrm{H}},B_{\mathrm{L}})} &= \mathrm{P}^{ch}_{n,k}\mathrm{P}^{tr}_{i,j}\mathrm{P}_{bl}\left(B_{\mathrm{H}} - 1|\lambda_{\mathrm{H},i}\right) \\ &\times \mathrm{P}_{bl}\left(B_{\mathrm{L}} - 1 - \max(0, q_{\mathrm{L}} - K_n)|\lambda_{\mathrm{L}}\right).\end{aligned} \tag{11.28}$$

11.3.3 System Performance Metrics

Based on the transition matrix $\mathbf{P}^s$ obtained in Sections 11.3.1 and 11.3.2, system performance metrics such as PDR, PER, throughput, and delay can be derived in closed-form expressions. This requires solving for the steady-state probabilities of $\mathbf{P}^s$. Let $(N_{\mathrm{AMC}} + 1)\, N_{\mathrm{H}}\, (B_{\mathrm{H}} + 1)(B_{\mathrm{L}} + 1)$ denote the size of this matrix, i.e., the total number of states in the Markov chain. The steady-state probability vector π^s can then be obtained from the following system of linear equations:

$$\begin{cases}\pi^s \mathbf{P}^s = \pi^s, \\ \displaystyle\sum_{n=0}^{N_{\mathrm{AMC}}} \sum_{i=1}^{N_{\mathrm{H}}} \sum_{q_{\mathrm{H}}=0}^{B_{\mathrm{H}}} \sum_{q_{\mathrm{L}}=0}^{B_{\mathrm{L}}} \pi^s_{(n,i,q_{\mathrm{H}},q_{\mathrm{L}})} = 1.\end{cases} \tag{11.29}$$

The steady-state probability vector π^s can be obtained by solving the above system of linear equations using standard numerical techniques, such as Gauss elimination or Jacobi iteration, and is given as

$$\begin{aligned}\pi^s = \Big[&\pi^s_{(1,1,0,0)}, \pi^s_{(1,1,0,1)}, \dots, \pi^s_{(n,i,q_{\mathrm{H}},q_{\mathrm{L}})}, \dots, \\ &\pi^s_{(N_{\mathrm{AMC}}+1,N_{\mathrm{H}},B_{\mathrm{H}}+1,B_{\mathrm{L}}+1)}\Big],\end{aligned} \tag{11.30}$$

which in turn facilitates the derivation of the system performance metrics for the considered CLD, as detailed in the following sections.

11.3.3.1 PDR

Obviously, packets arriving at a full buffer are dropped owing to buffer overflow. To evaluate the PDR, we first compute the average arrival rates of the two traffic classes. For the non-real-time traffic, the average number of arriving packets can be expressed as

$$\overline{T}_{\mathrm{L}} = \mathbb{E}\{q_{\mathrm{L}}\} = \lambda_{\mathrm{L}} T_f. \tag{11.31}$$

For real-time traffic modelled as a multistate MMPP, the average arrival rate can be expressed as follows. Let $\pi\mathsf{H} = [\pi\mathsf{H}_1, \pi\mathsf{H}_2, \ldots, \pi\mathsf{H}_{N_\mathsf{H}}]$ denote the steady-state probability vector of the N_H-state Markov chain. The vector is computed by solving the following system of linear equations

$$\begin{cases} \pi_\mathsf{H} = \pi_\mathsf{H}\mathbf{P}^{tr}, \\ \sum_{i=1}^{N_\mathsf{H}} \pi_{\mathsf{H}_i} = 1. \end{cases} \tag{11.32}$$

The average number of packets arriving for real-time traffic is

$$\overline{T}_\mathsf{H} = \sum_{i=1}^{N_\mathsf{H}} \lambda_{\mathsf{H},i}\pi_{\mathsf{H}_i}. \tag{11.33}$$

So, the PDR for both traffic classes, P_d^A, $\mathsf{A} \in \{\mathsf{H}, \mathsf{L}\}$, is computed as

$$P_d^\mathsf{A} = \zeta_d^\mathsf{A}/\overline{T}_\mathsf{A}, \tag{11.34}$$

where ζ_d^A, the average number of packets dropped during a time slot, is computed separately for high- and low-priority data.

First, we investigate the case of high-priority data. Let q_H denote the number of packets in the buffer at a given time slot, if the number of arriving packets exceeds $B_\mathsf{H} - q_\mathsf{H}$, buffer overflow occurs. In other words, when $n = 0$, the number of packets dropped can be expressed as $u\mathsf{P}_{ar}\left(u + B_\mathsf{H} - q_\mathsf{H}|\lambda_{\mathsf{H},i}\right)$, where $u = 1, 2, \ldots, \infty$ is the number of packets dropped during a time slot. However, when the channel is not in outage ($n \neq 0$), the computation slightly differs. After transmission, the number of packets in the buffer becomes $\max(0, q_\mathsf{H} - K_n)$. So, the number of packets dropped is $u\mathsf{P}_{ar}\left(u + B_\mathsf{H} - \max(0, q_\mathsf{H} - K_n)|\lambda_{\mathsf{H},i}\right)$. By combining the two cases ($n = 0$ and $n \neq 0$), the average number of packets dropped for high-priority data, ζ_d^H, can be expressed as

$$\begin{aligned} \zeta_d^\mathsf{H} = {} & \sum_{n=1}^{N_\mathsf{AMC}} \sum_{i=1}^{N_\mathsf{H}} \sum_{q_\mathsf{H}=0}^{B_\mathsf{H}} \sum_{q_\mathsf{L}=0}^{B_\mathsf{L}} \pi^s_{(n,i,q_\mathsf{H},q_\mathsf{L})} \\ & \times \sum_{u=1}^{\infty} u\mathsf{P}_{ar}\left(u + B_\mathsf{H} - \max(0, q_\mathsf{H} - K_n)|\lambda_{\mathsf{H},i}\right) \\ & + \sum_{i=1}^{N_\mathsf{H}} \sum_{q_\mathsf{H}=0}^{B_\mathsf{H}} \sum_{q_\mathsf{L}=0}^{B_\mathsf{L}} \pi^s_{(0,i,q_\mathsf{H},q_\mathsf{L})} \sum_{u=1}^{\infty} u\mathsf{P}_{ar}\left(u + B_\mathsf{H} - q_\mathsf{H}|\lambda_{\mathsf{H},i}\right). \end{aligned} \tag{11.35}$$

Following a similar reasoning, the average number of packets dropped for low-priority data can be obtained as follows:

$$\begin{aligned}\zeta_d^{\mathrm{L}} =& \sum_{n=1}^{N_{\mathrm{AMC}}} \sum_{i=1}^{N_{\mathrm{H}}} \sum_{q_{\mathrm{H}}=0}^{K_n-1} \sum_{q_{\mathrm{L}}=0}^{B_{\mathrm{L}}} \pi^s_{(n,i,q_{\mathrm{H}},q_{\mathrm{L}})} \\ &\times \sum_{u=1}^{\infty} u\mathrm{P}_{ar}\left(u + B_{\mathrm{L}} - \max(0, q_{\mathrm{L}} + q_{\mathrm{H}} - K_n)|\lambda_{\mathrm{L}}\right) \\ &+ \sum_{n=1}^{N_{\mathrm{AMC}}} \sum_{i=1}^{N_{\mathrm{H}}} \sum_{q_{\mathrm{H}}=K_n}^{B_{\mathrm{H}}} \sum_{q_{\mathrm{L}}=0}^{B_{\mathrm{L}}} \pi^s_{(n,i,q_{\mathrm{H}},q_{\mathrm{L}})} \sum_{u=1}^{\infty} u\mathrm{P}_{ar}\left(u + B_{\mathrm{L}} - q_{\mathrm{L}}|\lambda_{\mathrm{L}}\right). \\ &+ \sum_{i=1}^{N_{\mathrm{H}}} \sum_{q_{\mathrm{H}}=0}^{B_{\mathrm{H}}} \sum_{q_{\mathrm{L}}=0}^{B_{\mathrm{L}}} \pi^s_{(0,i,q_{\mathrm{H}},q_{\mathrm{L}})} \sum_{u=1}^{\infty} u\mathrm{P}_{ar}\left(u + B_{\mathrm{L}} - q_{\mathrm{L}}|\lambda_{\mathrm{L}}\right).\end{aligned} \tag{11.36}$$

The first summand in (11.36) corresponds to Case 5 in Section 11.3, where the low-priority data is readily delivered and the channel is not in outage ($n \neq 0$). In this case, the buffer-L overflows if the number of arriving low-priority packets exceeds $B_{\mathrm{L}} - \max(0, q_{\mathrm{L}} + q_{\mathrm{H}} - K_n)$. The second summand accounts for the scenario in which only high-priority data is allowed to transmit. Finally, the third summand corresponds to the outage case, i.e., $n = 0$.

11.3.3.2 PER

Due to the uncertainty of the time-varying channel, packets may be received with errors at the destination. This type of loss in service integrity is quantified by PER, defined as $P_e^{\mathrm{A}} = \mu_e^{\mathrm{A}}/\overline{T}_{\mathrm{A}}$, where μ_e^{A} denotes the average number of packets lost due to transmission errors, given by

$$\mu_e^{\mathrm{H}} = \sum_{n=1}^{N_{\mathrm{AMC}}} \sum_{i=1}^{N_{\mathrm{H}}} \sum_{q_{\mathrm{H}}=0}^{B_{\mathrm{H}}} \sum_{q_{\mathrm{L}}=0}^{B_{\mathrm{L}}} \pi^s_{(n,i,q_{\mathrm{H}},q_{\mathrm{L}})} \min(q_{\mathrm{H}}, K_n)\overline{\mathrm{PER}}_n, \tag{11.37}$$

$$\mu_e^{\mathrm{L}} = \sum_{n=1}^{N_{\mathrm{AMC}}} \sum_{i=1}^{N_{\mathrm{H}}} \sum_{q_{\mathrm{H}}=0}^{K_n-1} \sum_{q_{\mathrm{L}}=0}^{B_{\mathrm{L}}} \pi^s_{(n,i,q_{\mathrm{H}},q_{\mathrm{L}})} \min(q_{\mathrm{L}}, K_n - q_{\mathrm{H}})\overline{\mathrm{PER}}_n. \tag{11.38}$$

11.3.3.3 Throughput and Delay

The packet loss rate of each traffic class comprises both PDR and PER. Hence, the throughput is

$$\eta^{\mathrm{A}} = \left(1 - P_d^{\mathrm{A}} - P_e^{\mathrm{A}}\right)\overline{T}_{\mathrm{A}}. \tag{11.39}$$

Delay, which is an important system performance metric, is written as

$$D^{\mathsf{A}} = D^{\mathsf{A}}_{\mathsf{ETI}} + D^{\mathsf{A}}_{\mathsf{Q}}, \tag{11.40}$$

where $D^{\mathsf{A}}_{\mathsf{ETI}} = T_f/2$ denotes the delay incurred by the enable transmission interval (ETI), and $D^{\mathsf{A}}_{\mathsf{Q}}$ is the average queuing delay, given by the ratio of the average number of queuing packets Q^{A} to the effective packet arrival rate $(1 - P^{\mathsf{A}}_d)\overline{T}_{\mathsf{A}}$, expressed as

$$D^{\mathsf{A}}_{\mathsf{Q}} = \frac{Q^{\mathsf{A}}}{(1 - P^{\mathsf{A}}_d)\overline{T}_{\mathsf{A}}}, \tag{11.41}$$

where Q^{A} can be easily obtained as

$$Q^{\mathsf{A}} = \sum_{n=0}^{N_{\mathsf{AMC}}} \sum_{i=1}^{N_{\mathsf{H}}} \sum_{q_{\mathsf{H}}=0}^{B_{\mathsf{H}}} \sum_{q_{\mathsf{L}}=0}^{B_{\mathsf{L}}} \pi^{s}_{(n,i,q_{\mathsf{H}},q_{\mathsf{L}})} q_{\mathsf{A}}. \tag{11.42}$$

11.4 CLD Problem and Solution

We have thoroughly discussed the performance metrics of the CLD in closed-form expressions. In this section, we utilize these expressions to formulate an optimization problem for general MAS. We then apply the proposed CLD optimization to a specific case of video streaming.

11.4.1 Optimization Problem for MAS

In this scenario, the two traffic classes are categorized into real-time (e.g., mobile video applications) and non-real-time (e.g., data downloading). As observed from (11.39) and (11.41), throughput and delay also depend on the packet loss rate. Consequently, our objective is to minimize $(P^{\mathsf{A}}_d + P^{\mathsf{A}}_e)$. To this end, we formulate constrained linear optimization problems with two corresponding objective functions. Solving these problems yields the optimal values of B_{L}, B_{H}, and $\overline{\mathrm{PER}}_n$, thereby minimizing the overall packet loss rate.

11.4.1.1 Minimum PDR

According to the relationship between P^{A}_d and ζ^{A}_d, minimizing $(P^{\mathsf{L}}_d + P^{\mathsf{H}}_d)$ is equivalent to minimizing the total average number of dropped packets, i.e., $(\zeta^{\mathsf{L}}_d + \zeta^{\mathsf{H}}_d)$, for both real-time and non-real-time traffic. The optimization problem can thus be formulated as follows:

$$\min_{B_{\mathsf{L}},B_{\mathsf{H}}} (\zeta^{\mathsf{L}}_d + \zeta^{\mathsf{H}}_d) \tag{11.43}$$

$$\text{s.t.} \begin{cases} \dfrac{\zeta_d^{\mathsf{L}}}{\zeta_d^{\mathsf{H}}} \leq \alpha, \\ \dfrac{D_Q^{\mathsf{L}}}{D_Q^{\mathsf{H}}} \leq \beta, \end{cases} \tag{11.44}$$

where (11.44) is used to limit the PDR and the delay of non-real-time traffic by the coefficients $\alpha \geq 1$ and $\beta \geq 1$.

11.4.1.2 Minimum PER

Similarly, in order to minimize P_e^{A}, it is necessary to minimize the total average number of packets lost due to transmission errors across both low- and high-priority traffic classes, i.e., $(\mu_e^{\mathsf{L}} + \mu_e^{\mathsf{H}})$. The corresponding optimization problem can be formulated as follows:

$$\min_{\overline{\text{PER}_n}} (\mu_e^{\mathsf{L}} + \mu_e^{\mathsf{H}}) \tag{11.45}$$

$$\text{s.t.} \begin{cases} \sum\limits_{n=1}^{N_{\mathsf{AMC}}} \overline{\text{PER}}_n = N_{\mathsf{AMC}} P_0, \\ 0 < \overline{\text{PER}}_n \leq 1. \end{cases} \tag{11.46}$$

In the above optimization problem, it is important to note that determining the optimal set $\{\overline{\text{PER}}_n\}_{n=0}^{N_{\mathsf{AMC}}}$ is equivalent to computing the optimal switching thresholds $\{\gamma_n\}_{n=0}^{N_{\mathsf{AMC}}}$. However, directly obtaining γ_n is not trivial. Instead, we first derive the optimal set $\{\overline{\text{PER}}_n\}_{n=0}^{N_{\mathsf{AMC}}}$ and then compute γ_n by solving the following equations:

$$\begin{cases} N_{\mathsf{AMC}} P_0 = \sum\limits_{n=1}^{N_{\mathsf{AMC}}} \dfrac{1}{\int_{\gamma_n}^{\gamma_{n+1}} f_\gamma(\gamma)\, d\gamma} \displaystyle\int_{\gamma_n}^{\gamma_{n+1}} a_n \exp(-g_n \gamma) f_\gamma(\gamma)\, d\gamma, \\ \overline{\text{PER}}_n = \dfrac{1}{\int_{\gamma_n}^{\gamma_{n+1}} f_\gamma(\gamma)\, d\gamma} \displaystyle\int_{\gamma_n}^{\gamma_{n+1}} a_n \exp(-g_n \gamma) f_\gamma(\gamma)\, d\gamma. \end{cases} \tag{11.47}$$

11.4.2 Optimization Problem for Video Streaming

To evaluate the efficiency of the proposed CLD, we consider a representative case of MAS in which video content is delivered to MUs. In this chapter, we assume that fine granularity scalability (FGS) encoded sequences based on the MPEG-4 standard are employed [17, 18]. An FGS-encoded sequence consists of two streams: a non-scalable base layer (BL) providing basic reconstruction quality and a scalable enhancement layer (EL) providing additional refinement. Since BL data plays a critical role in the reconstructed quality of video streaming, it is treated as real-time traffic with high-priority transmission, while EL data is classified as non-real-time traffic.

In this context, our objective is to minimize the average reconstructed distortion of the received video. This distortion stems from source encoding, transmission congestion, and transmission errors, represented by both PDR and PER. Consider a video stream with M frames, where the source distortion of frame k is given by $d_k = d_k^{\mathrm{BL}} - d_k^{\mathrm{EL}}(r_k)$. Here, d_k^{BL} denotes the distortion when only the BL frame is reconstructed (i.e., without enhancement), and $d_k^{\mathrm{EL}}(r_k)$ represents the reduction in distortion if the EL frame is encoded at rate r_k. Let BL frame k consist of n_k^{BL} packets and EL frame k consist of n_k^{EL} packets. Owing to their small size and critical importance, BL frames are protected by forward error correction and are always correctly reconstructed at the receiver [19, 20]. By accounting for transmission congestion and errors, the reconstructed distortion of frame k at the receiver can be expressed as

$$d_k = d_k^{\mathrm{BL}} - [(1 - P_d^{\mathrm{H}})(1 - P_e^{\mathrm{H}})]^{n_k^{\mathrm{EL}}} d_k^{\mathrm{EL}}(r_k). \tag{11.48}$$

Accordingly, the average reconstructed distortion of the received video is computed as

$$\overline{d} = \frac{1}{M} \sum_{k=1}^{M} d_k. \tag{11.49}$$

Consequently, (11.49) serves as the objective function in the optimization problem of video streaming applications, which can be formulated as follows:

$$\min_{B_{\mathrm{L}}, B_{\mathrm{H}}, \overline{\mathrm{PER}}_n} \overline{d} \tag{11.50}$$

$$\text{s.t.} \begin{cases} B_{\mathrm{H}} + B_{\mathrm{L}} = B_{\mathrm{tot}}, \\ \sum_{n=1}^{N_{\mathrm{AMC}}} \overline{\mathrm{PER}}_n = N_{\mathrm{AMC}} P_0, \\ 0 < \overline{\mathrm{PER}}_n \leq 1 \\ D_{\mathrm{Q}}^{\mathrm{H}} \leq D_{\mathrm{Q}}^{\mathrm{th}}, \\ |d_{k+1} - d_k| \leq \delta, \end{cases} \tag{11.51}$$

where B_{tot} denotes the total buffer size. The fourth constraint ensures continuous playback by the targeted delay $D_{\mathrm{Q}}^{\mathrm{th}}$. This constraint is critical since video streaming is highly delay-sensitive. The last constraint, with a threshold δ, guarantees smooth playback of the received video. Its necessity arises from the fact that multiple frames of a video are often transmitted using different AMC modes at the PHY layer, leading to highly skewed reconstructed distortions. Such skewness hinders smooth playback, and thus limiting it is essential for robust and efficient optimization.

11.5 Performance Evaluation of CLD

In this section, we present numerical results to evaluate the impact of system parameters on the performance of the proposed CLD. We first validate the analytical results by comparing them with Monte Carlo simulations. Then, we demonstrate the benefits of CLD optimization in two representative scenarios: (i) general multimedia traffic with high- and low-priority classes and (ii) video streaming over wireless networks.

11.5.1 CLD for MAS

We set the packet length to $N_p = 1080$ bits and the frame duration to $T_f = 2$ ms. The Doppler frequency is chosen as $f_d = 15$ Hz. For the considered Nakagami-m fading channel, we use the fading severity parameter $m = 2$ and the number of transmission modes $N_{\mathrm{AMC}} = 4$. The target PER is $P_0 = 0.01$. The total buffer size is fixed at $B = 12$ packets, equally partitioned between the two traffic classes with $B_{\mathrm{H}} = 6$ and $B_{\mathrm{L}} = 6$. For multimedia traffic, the arrival rate of non-real-time traffic is set to $\lambda_{\mathrm{L}} = 0.5\lambda$, where λ denotes the total packet arrival rate in kilo-packets per second (kpkts/s). To capture the multiple states of real-time traffic, the arrival rate vector is defined as $\boldsymbol{\lambda}^{\mathrm{H}} = \begin{bmatrix} \lambda & 0.1\lambda \end{bmatrix}$ and transition matrix is $\mathbf{P}^{tr} = \begin{bmatrix} 0.8 & 0.2 \\ 0.4 & 0.6 \end{bmatrix}$. Since the packet loss rate, i.e., the sum of PDR and PER, is already reflected in the delay and throughput metrics given in (11.41) and (11.39), we present only the delay and throughput results to avoid overly crowded figures.

11.5.1.1 Dependence on Arrival Rate

We present the performance metrics of the two traffic classes with respect to the packet arrival rate λ at a fixed average SNR of $\gamma_0 = 15$ dB. Figure 11.1a,b illustrate the service integrity performance in terms of delay and throughput versus λ. As observed, the analytical results in Section 11.3 closely match the simulation outcomes. Figure 11.1a demonstrates that real-time traffic achieves lower delay than non-real-time traffic. This is expected since real-time traffic is prioritized in transmission, which reduces the number of queued high-priority packets. Consequently, as shown in Figure 11.1b, high-priority traffic also attains higher throughput compared to low-priority traffic.

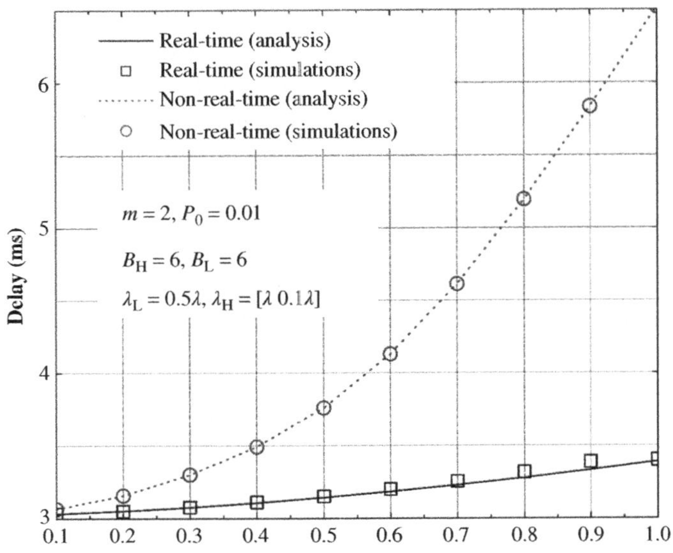

(a)

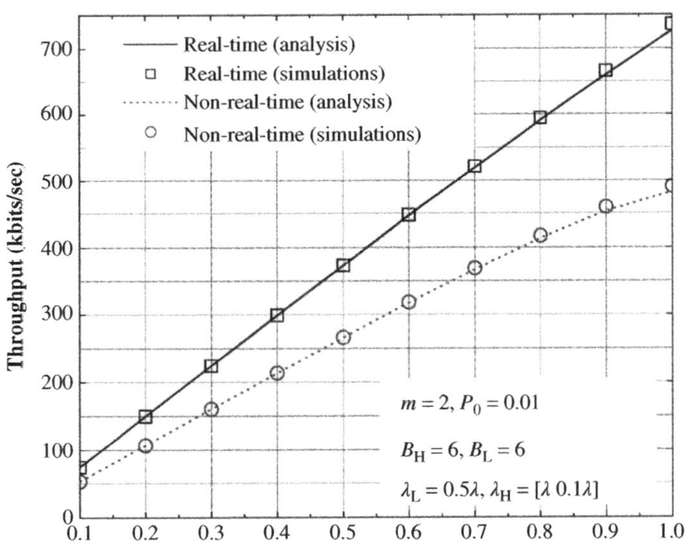

(b)

Figure 11.1 Performance of CLD versus packet arrival rate λ: comparison between real-time and non-real-time traffic classes. (a) Delay. (b) Throughput. *Source:* Reproduced from [1] / IEEE.

11.5.1.2 Dependence on Average SNR

To evaluate the dependence of the system performance on the average SNR γ_0, we fix the packet arrival rates for non-real-time and real-time traffic classes as $\lambda_L = 0.5$ kpkt/sec and $\lambda_H = [10.1]$ kbps, respectively, and varying γ_0 from 0 to 15 dB. The delay and throughput versus γ_0 are plotted in Figure 11.2a,b. Intuitively, we can see that the performance improves with the increase of γ_0. As can be observed from the two figures, the performance in the low SNR regime is very poor since the low-rate transmission modes are frequently selected. For example, when $\gamma_0 = 15$ dB, a significant increase of approximately 200 kbps can be obtained as compared to the case $\gamma_0 = 5$ dB. As expected, the real-time traffic which is very sensitive to delay, communicates with much shorter average delay than non-real-time one. In addition, real-time traffic results in higher throughput than non-real-time one for the whole considered SNR range.

11.5.2 CLD for Video Streaming

From the constrained optimization problem in (11.43) and (11.44), the overall PDR is minimized when B_L and B_H converge to five under the constraint $B_{tot} = 10$. These optimal values of B_L and B_H are subsequently applied to the constrained optimization problem in (11.45) and (11.46) to determine the optimal values of $\overline{PER_n}$. Solving this problem yields the minimum PER. Clearly, the proposed method achieves lower PDR and PER, leading to higher throughput compared with the non-optimized approach, as illustrated in Figure 11.3. The system parameters for this scenario differ from those in Section 11.4.1, with the fading severity parameter set to $m = 1$, the average SNR is fixed at $\gamma_0 = 15$ dB, and the state transition matrix for real-time traffic is $\mathbf{P}^{tr} = \begin{bmatrix} 0.9 & 0.1 \\ 0.3 & 0.7 \end{bmatrix}$.

We employ the proposed CLD to optimize the performance of video streaming. The Foreman sequence is analyzed using the MPEG-4 Reference Software Version 2.5 [21], supported by Microsoft, with CIF resolution (352×288), 30 frames per second, and 6 bit-planes. Due to the diverse bandwidth requirements of multiple MUs, the video may not be encoded at the full encoding rate. This strategy is commonly employed in dense networks to conserve the bandwidth. Consequently, the MUs often request the video at a receive bandwidth corresponding to approximately 70% of the full encoding rate, i.e., the peak signal-to-noise ratio (PSNR) of the Foreman video is equivalent to 49.82 dB. To this end, the rate–distortion model (RDM) of each frame is generated following the method in [20] with the assistance of [21]. For each RD curve, the frame distortion is truncated and measured at 70% of the full encoding rate. By applying the same procedure to all RD curves, the average distortion is computed and then converted into the equivalent PSNR value. To evaluate the performance of video streaming applications, we consider three key

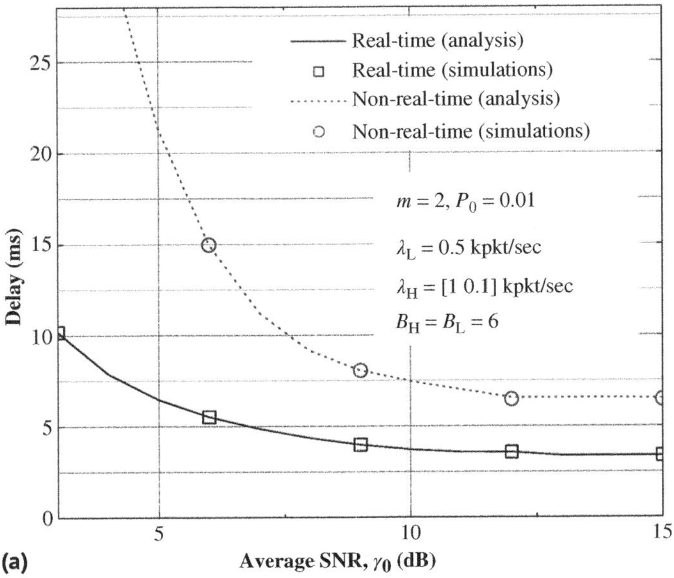

(a)

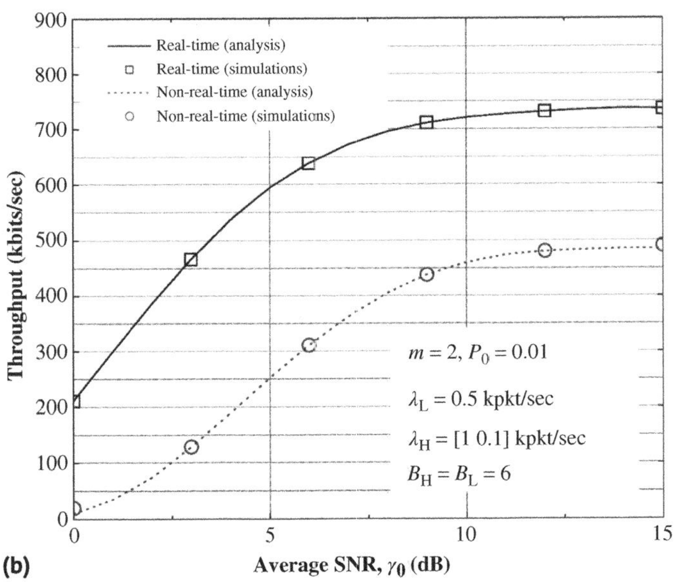

(b)

Figure 11.2 Performance of CLD versus average SNR γ_0: comparison between real-time and non-real-time traffic classes. (a) Delay. (b) Throughput. *Source:* Reproduced from [1] / IEEE.

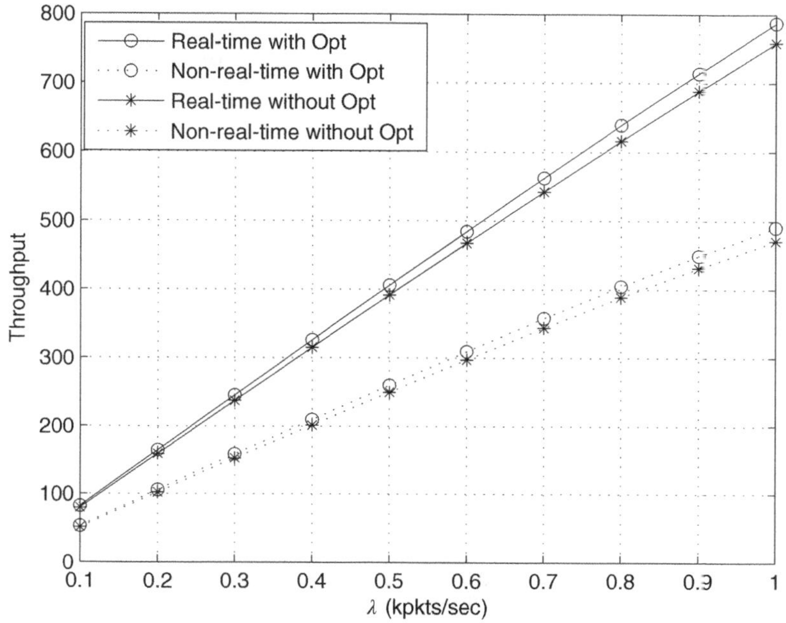

Figure 11.3 Throughput of real-time and non-real-time data versus λ with $N_{\mathrm{AMC}} = 6$. *Source:* Reproduced from [1] / IEEE.

metrics directly related to the QoE perceived by the MUs: minimum reconstructed distortion for high-quality playback, continuous playback, and smooth playback [19]. With respect to the constraints in (11.51), we set the following parameters: $B_{\mathrm{tot}} = 10$, $N_{\mathrm{AMC}} = 4$, $P_0 = 0.03$, and $D_{\mathrm{Q}}^{\mathrm{th}} = 4.5$ ms. In addition, the smooth playback threshold δ in the last constraint of (11.51) is chosen as 10% of the total distortion of the original video, i.e.,

$$\delta = 10\% \sum_{k=1}^{M} \left[d_k^{\mathrm{BL}} - d_k^{\mathrm{EL}}(r_k)\right]. \tag{11.52}$$

For quality playback evaluation, we employ the PSNR metric, which is calculated from the mean squared error (MSE) between the original and reconstructed video sequences as $\mathrm{PSNR} = 10 \log_{10}\left(\frac{255^2}{\mathrm{MSE}}\right)$. As illustrated in Figure 11.4, CLD optimization enhances the PSNR. However, its advantage diminishes when the packet arrival rate λ becomes excessively high. A higher arrival rate leads to a larger PDR, and thus lower perceived quality. Specifically, when λ exceeds the buffer capacity B (with $B = 10$, $\lambda = 1$ kpkt/s is already too large), buffer overflow occurs, rendering the preemptive priority mechanism ineffective. Consequently, the performance of CLD optimization degrades and gradually converges to that of

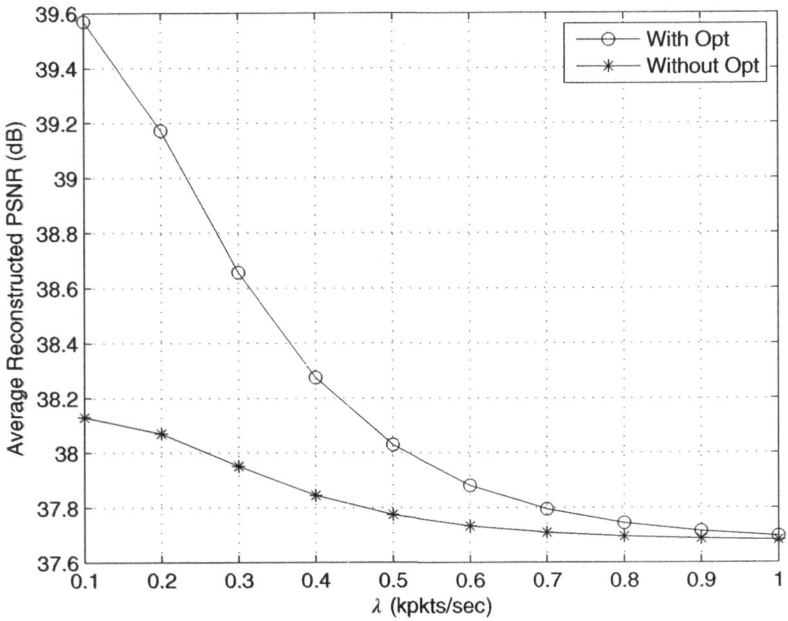

Figure 11.4 PSNR performance versus λ with $N_{\text{AMC}} = 4$, $B_{\text{H}} = 4$, and $B_{\text{L}} = 6$.
Source: Reproduced from [1] / IEEE.

the non-optimized scheme as λ increases. This highlights the critical role of incorporating finite-buffer queuing analysis into CLD optimization.

To evaluate the smooth playback, we compute the standard deviation of the PSNR across received frames as

$$\sigma_d = \sqrt{\frac{1}{M}\sum_{k=1}^{N}(d_{\text{B}k} - \bar{d}_{\text{B}})^2}, \tag{11.53}$$

where $d_{\text{B}k} = 10\log_{10}\left(255^2/d_k\right)$ and $\bar{d}_{\text{B}} = 10\log_{10}\left(255^2/\bar{d}\right)$. As shown in Figure 11.5, the proposed optimal method achieves a smaller PSNR standard deviation than the non-optimal approach, indicating improved playback smoothness. Interestingly, in addition to enhancing video quality and smoothness, our method also meets the continuous-playback delay constraint. Specifically, Figure 11.6 shows that the streaming delay remains below the target $D_{\text{Q}}^{\text{th}} = 4.5$ ms across the entire considered range of packet arrival rates λ.

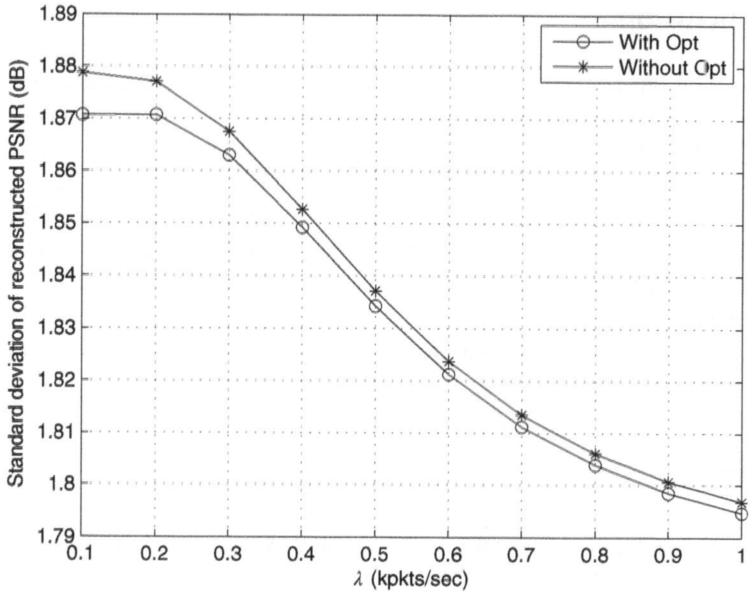

Figure 11.5 Smooth playback performance versus λ with $N_{AMC} = 4$, $B_H = 4$, and $B_L = 6$. *Source:* Reproduced from [1] / IEEE.

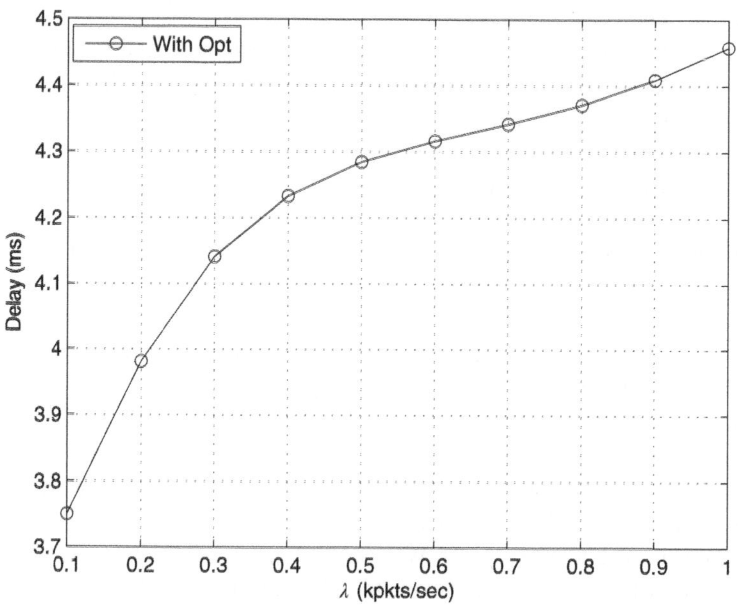

Figure 11.6 Continuous playback constraint represented by delay versus λ with $N_{AMC} = 4$, $B_H = 4$, and $B_L = 6$. *Source:* Reproduced from [1] / IEEE.

11.6 Conclusion

CLD optimization plays a key role in QoS and QoE provisioning. We have proposed a general framework for CLD in MAS over CDN, integrating QoS and QoE guarantees for real-time and non-real-time traffic classes with buffer scheduling at the APP and DLK layers, and AMC at the PHY layer. The CLD was modelled as a four-dimensional Markov chain, enabling closed-form expressions for PDR, PER, delay, and throughput. This analysis can be leveraged to optimize various MASs. In particular, our CLD enhanced video streaming over wireless networks, demonstrating improved system performance compared to a non-optimized approach. Moreover, the scheme preserves critical QoE metrics, such as playback quality and smoothness in video delivery.

References

1 N.-S. Vo, T.Q. Duong, H.-J. Zepernick, and M. Fiedler. A cross-layer optimized scheme and its application in mobile multimedia networks with QoS provision. *IEEE Syst. J.*, 10(2):817–830. 2014.

2 G. Pan, S. Xu, S. Zhang, X. Chen, and Y. Sun. Quality of experience oriented cross-layer optimization for real-time XR video transmission. *IEEE Trans. Circuits Syst. Video Technol.*, 34(8):7742–7755. 2024.

3 S.-M. Tseng, S.-T. Wen, C. Fang, and M. Norouzi. Cross layer power allocation by graph neural networks in heterogeneous D2D video communications. *IEEE Access*, 13:44484–44496. 2025.

4 T.-N. Tran and B. An. QoS multicast routing utilizing cross-layer design for IoT-enabled MANET in RIS-aided cell-free massive MIMO. *IEEE Internet Things J.*, 11(7):11876–11893. 2024.

5 Q. Wu, H. Zhuang, and L. Zhang. SON-RRM intelligent optimization based on cross-layer deep reinforcement learning. In *Proc. IEEE Veh. Technol. Conf.*, pages 1–6, Washington, DC, Oct. 2024.

6 L.-W. Chien and S.-M. Tseng. Proximal policy optimization for cross-layer joint design of MISO beamforming and RIS phase configuration. *IEEE Open J. Commun. Soc.*, 6:8128–8141. 2025.

7 M. Takahashi and A. Nakao. Cross-layer congestion control algorithm and architecture for communication performance optimization in millimeter-wave networks. *IEEE Access*, 14:35823–35837. 2025. Early Access.

8 Y. Zhang, W. Han, Y. Chong, X. Ma, W. Wu, and W. Teng. Cross-layer optimization of scalable video transmission for IoV in 6G. In *IEEE/CIC Inter. Conf. Commun. in China*, pages 1–6, Hangzhou, China, Aug. 2024.

9 F. Feng, G. Liu, T. Bi, and T. Jiang. Edge selective sharing for massive mobile video streaming with cross-layer optimization. *IEEE Trans. Mob. Comput.*, 23(12):11907–11919. 2024.

10 E. M. Ghourab, M. Azab, D. Gračanin, M. Al-Qutayri, and S. Muhaidat. Cross-layer management framework for enhancing XR-based system security in zero-trust wireless communications. *IEEE J. Sel. Areas Commun.*, 43(6):2011–2024. 2025.

11 V. S. Frost and B. Melamed. Traffic modeling for telecommunications networks. *IEEE Commun. Mag.*, 32(3):70–81. 1994.

12 Q. Liu, S. Zhou, and G. B. Giannakis. Queuing with adaptive modulation and coding over wireless links: Cross-layer analysis and design. *IEEE Trans. Wireless Commun.*, 4(3):1142–1153. 2005.

13 Q. Liu, S. Zhou, and G. B. Giannakis. Cross-layer combining of adaptive modulation and coding with truncated ARQ over wireless links. *IEEE Trans. Wireless Commun.*, 3(5):1746–1755. 2004.

14 Q. Liu, S. Zhou, and G. B. Giannakis. Cross-layer scheduling with prescribed QoS guarantees in adaptive wireless networks. *IEEE J. Sel. Areas Commun.*, 23(5):1056–1066. 2004.

15 Q. Liu, X. Wang, and G. B. Giannakis. A cross-layer scheduling algorithm with QoS support in wireless networks. *IEEE Trans. Veh. Technol.*, 55(3):839–847. 2006.

16 H. S. Wang and N. Moayeri. Finite-state Markov channel - A useful model for radio communication channels. *IEEE Trans. Veh. Technol.*, 44(1):163–171. 1995.

17 W. Li. Overview of fine granularity scalability in MPEG-4 video standard. *IEEE Trans. Circuits Syst. Video Technol.*, 11(3):301–317. 2001.

18 H. Radha, M. V. D. Schdar, and Y. Chen. The MPEG-4 fine-grained scalable video coding method for multimedia streaming over IP. *IEEE Trans. Multimed.*, 3(1):53–68. 2001.

19 N.-S. Vo, T. Q. Duong, L. Shu, X. Du, H.-J. Zepernick, and W. Cheng. Cross-layer design for video replication strategy over multihop wireless networks. In *Proc. IEEE Inter. Commun. Conf.*, pages 1–6, Kyoto, Japan, Jun. 2011.

20 C. Hsu and M. Hefeeda. Optimal bit allocation for fine-grained scalable video sequences in distributed streaming environments. In *Proc. ACM/SPIE Multimedia Computing and Networking Conf.*, pages 1–12, San Jose, CA, Jan. 2007.

21 Reference Software. MPEG-4 visual reference software. In *ISO/IEC 14496*. 2004.

Index

n

o

p

q

r

s

Printed and bound by CPI Group (UK) Ltd, Croydon, CR0 4YY
17/09/2026
14964312-0001